Sir Solomon de Medina

Sir Solomon de Medina

by

Oskar K. Rabinowicz

and

a Biography of the author
by Judith K. Tapiero and Theodore K. Rabb

Published by
THE JEWISH HISTORICAL SOCIETY OF ENGLAND
33 SEYMOUR PLACE, LONDON W1
1974

© Oskar K. Rabinowicz (*Sir Solomon de Medina*)

© Judith K. Tapiero and Theodore Rabb (*Oskar K. Rabinowicz: a biographical sketch*)

Printed in Great Britain by
Alden & Mowbray Ltd
at the Alden Press, Oxford

Contents

	Page
FOREWORD	vii
INTRODUCTION	ix

Chapter
- *One:* Solomon de Medina: The data — 1
- *Two:* Medina's position in the Jewish community — 6
- *Three:* An important citizen — 14
- *Four:* Merchant — 23
- *Five:* Financier
  - (a) Bill discounting, credits, and loans — 26
  - (b) Financial speculation — 33
- *Six:* Army contractor — 37
- *Seven:* Wealth and decline — 43
- *Eight:* Medina and Marlborough — 61
-   Medina and Voltaire — 70

Appendix
- A: The identity of Solomon and Diego — 75
- B: Joseph de Medina's family — 82
- C: (a) The will of Sir Solomon and Lady de Medina — 94
-   (b) Lady de Medina's codicil — 107
- D: Solomon de Medina's answer to Moses de Medina — 113
- E: Moses Dias's dedication to Solomon de Medina — 116
- F: Glossary — 118

Oskar K. Rabinowicz: Biography — 121

INDEX—*Sir Solomon de Medina* — 143

INDEX—*Oskar K. Rabinowicz* — 152

Illustrations

	Page
Sir Solomon and Lady de Medina's tombstones, diagram	5
Solomon de Medina's signature	11
Partition Book recording the Knighthood (1700/1701)	21
Christopher Dodsworth's proceedings over exports of silver	78–79
Oskar K. Rabinowicz	*facing* 121

Foreword

This book is an extension (by the author) of a lecture delivered by Dr. Oskar K. Rabinowicz before the Jewish Historical Society of England on 19 June 1947. It has not been printed before because the author had always felt that he would like to do further research on the subject, in keeping with the meticulous care he had exercised in publishing his other historical works. In between other insistent and onerous duties in the U.S.A., Israel, and England, he nevertheless found time to make further visits to the Public Record Office in London, and expanded the work through additional research mainly in the Treasury State Papers, Domestic State Papers, and Chancery Court Proceedings. He was still engaged on such work when he so unfortunately died. It has been of course impossible to consult the author on some historical doubts, but in the resolving of these thankful acknowledgments should be made to Mrs. Judith Tapiero—mentioned below—and Dr. Aubrey Newman, Reader in History at Leicester University.

In addition to the paper on Solomon de Medina, the Jewish Historical Society of England has thought that it would not be inappropriate to include in the same volume a biography of the author, written as a memorial tribute to him after his greatly lamented death on 26 June 1969. The authors of this biography are his son, Professor Theodore Rabb, Professor of History at Princeton University, U.S.A., and his daughter, Mrs. Judith Tapiero, B.A., M.L.S. The Jewish Historical Society of England takes this opportunity of expressing its sincere gratitude to Mrs. Rose Rabinowicz, wife of the author, and to her son and daughter, and wishes to acknowledge the kind participation of the Society for the History of Czechoslovak Jews in the publication of this book.

JOHN M. SHAFTESLEY,
Editor of Publications

Introduction

On 23 June 1700 Solomon de Medina, one of the outstanding members of the Spanish and Portuguese Synagogue in London, was knighted by King William III at Hampton Court. One hundred and thirty-seven years were to elapse before the same dignity was to be conferred upon Moses Montefiore, then Sheriff of London, the next Jew to be so elevated and honoured. Indeed, it may be said that Solomon de Medina was the most prominent British Jew in the world of politics and business affairs down to the nineteenth century, and in his own day held a position perhaps unparalleled in all Europe. In fact, he was the only British Jew whose position can be compared with the great Court Jews of Central Europe at that time.

In English history generally, too, he plays a part of considerable importance, owing to his work in provisioning and financing the campaigns in the Low Countries before and during the War of the Spanish Succession and his close association with the Duke of Marlborough.

Insufficient attention has been paid to him hitherto by Anglo-Jewish historians, and the following study is the first attempt to have been made to give a consistent and detailed account of his remarkable and brilliant career.

O.K.R.

CHAPTER ONE

Solomon de Medina: The Data

Practically no evidence is available to fix Solomon de Medina's place and date of birth. Nor, curiously enough, are details of his parentage and origin stated in the sources. He was obviously of Marrano extraction, and as far as can be ascertained was Diego *alias* Solomon de Medina.[1]

His father was Francesco de Medina, who married Gracia Pereira,[2] a sister of Manuell Lopez Pereira. They lived in Bordeaux, and moved to Amsterdam[3] before 1656. Besides Solomon, they had an elder son, Joseph Aaron.[4]

Solomon was born in Bordeaux.[5] Only by way of implication can we calculate the approximate date of his birth. Wolf mentions[6] that 'before he [Solomon de Medina] was thirty years of age, he figures as the largest contributor to the funds of the Bevis Marks Congregation'. According to Picciotto,[7] Medina was already the 'largest contributor' to the Sephardi funds in 1677. Accepting Wolf's statement (the source for which the author did not reveal, nor have I been able to trace its authority), we assume that in 1677 Solomon de Medina was under thirty years of age, and was thus born approximately in 1650. This date receives further indirect confirmation in the

[1] See Appendix A.
[2] Lucien Wolf's papers deposited at University College London.
[3] *Cf.* Appendix B.
[4] *Cf.* Appendix B.
[5] Pat. Roll 24, Car. II. Pt. 4; Dr. Wm. A. Shaw, *Denizations and Naturalizations*, Lymington, 1911, p. 108 (Huguenot Society of London's Publications, XVIII). The denization grant, quoted in Annex A, refers to him and the others as 'natives of France'.
[6] Lucien Wolf, 'Queen Anne's Army Contractors', in *Jewish Chronicle*, 28 June 1889, p. 16.
[7] James Picciotto, *Sketches of Anglo-Jewish History*, London, 1875, p. 50. This same statement is made in an anonymous article, 'Jewish Princes of Finance—I. Solomon de Medina—The Lopez Family', in *Jewish World*, 1 February 1878, p. 6.

denization papers, which state[8] that Medina arrived in England with 'his wife and family' in 1672. Thus he was married when he was about twenty, which conforms with the custom among Sephardim of marrying at an early age.

Solomon de Medina was in his early childhood when he moved with his parents from Bordeaux to Amsterdam in 1656, where he then lived 'for several years'. He visited London frequently from Amsterdam,[9] and the Bevis Marks Synagogue records mention his presence there for the first time in 1670.[10] But it was only after the declaration of King Charles II of 12 June 1672

> for encouraging the subjects of the United Provinces of the Netherlands to transport themselves with their estates and to settle in this His Majesty's Kingdom of England[11]

that Diego applied for permission to settle in England. Unfortunately, this application, which might have revealed the missing data about his origin and birth date, has not been traceable. However, I was able to find the permission enabling him and his family to settle in England in an entry dated 21 June 1672, which constitutes the first official reference to Medina in the Government's files. This reads as follows:[12]

> The Treasury Lords to the Customs Commissioners to permit Diego de Medina, of Amsterdam, who desires to benefit of His Majesty's declaration [of the 12th instant] to transport all his goods and merchandise hither, paying only native customs; he and his family according to the said declaration being to receive all privileges and freedoms of a natural born subject.

He arrived with his family and goods some time between 21 June and 23 September 1672, as the following document confirms:[13]

[8] *Cf.* Appendix A.
[9] Denization application in Appendix A.
[10] *El Libro de los Acuerdos: being the record and accompts of the Spanish and Portuguese Synagogue from 1663–1681* (transl. from the Spanish by L. D. Barnett, 1931), p. 47.
[11] *S.P.Dom.* Vol. 71, p. 210.
[12] *S.P.Tr.* G.D. 32/48, p. 39; Vol. III, Pt. 2 (1669–1672), p. 1257 (Index).
[13] *S.P.Tr.* G.D. 32/48. p. 146; Vol. III, Pt. 2 (1669–1672), p. 1313.

The Treasury Lords to the Customs Commissioners. By our letter of June 21 last (made on the desire of Diego de Medina for the benefit of the King's declaration for transporting himself and his family with his goods and merchandises to London from Amsterdam) you were directed to permit him to bring hither all his goods and merchandises, paying only native's duty. He has accordingly so transported himself. 'According to His Majesty's gracious declaration he is entitled to all privileges and freedoms of any of His Majesty's naturall born subjects.' You are therefore to permit him as well to export as to import any goods and merchandises, paying the same customs that are paid by Englishmen.

Medina's wife was Ester d'Azevedo, and their marriage probably took place in Amsterdam, where, as already noted, Medina resided before moving to London in 1672. On his marriage he entered into a marriage settlement—Ketiba— amounting to fifteen thousand guilders,[14] or about £1,500.

They had a daughter, Deborah,[15] who was born before her parents' settlement in England. As quoted above, the denization papers refer to Medina's 'wife and family', which could indicate more than the one daughter. But there is no evidence traceable of another child. In some of Marlborough's correspondence a son of Medina's is mentioned;[16] a number of official documents also refer to a son of his.[17] But from the context it is quite obvious that it is not his son but his son-in-law, Moses de Medina, who is meant in all these references.

Medina and his family lived at first in Great St. Helens, in the City of London. *The Little London Directory of 1677*[18] records him as

Solo Demodina Great St. Hel-.

The Medina family moved to this home fairly soon after their arrival in London. The Guildhall Records list them there already in 1673/4.[19] Later he occupied a house in Richmond, which he may have chosen for health reasons, because according

[14] Medina's will (Appendix C).
[15] *Ibid.* See also Appendix D.
[16] *The Marlborough Dispatches* (ed. Sir George Murray), London, 1845, IV, p. 707; V, pp. 256 and 274.
[17] *S.P.Tr.* Treasure Minute Book XIII, p. 124; XVIII, pp. 183/184; Reference Book VIII, p. 426.
[18] *The Oldest Printed List of Merchants and Bankers of London.* Reprinted London, 1863 (the reference to Medina to be found under 'D').
[19] Guildhall Records Assessments, January 1673/4, Box 35, MS. 7.

to a contemporary document[20] at that time Richmond was

> a populous place and much resorted to for the benefit of the Air and the Pleasantness of its Situation.

The records do not reveal when Medina acquired the lease in Richmond. But he already lived there in 1699, when William III visited him at his house.[21] On leaving England in 1702,[22] Medina's son-in-law probably moved there, although the first reference to its occupation by Moses de Medina is dated 18 October 1722.[23] From this reference we get a description of the house, which was

> situated almost in the middle of the Town upon the first rise of Richmond Hill in a row of five houses leading up the hill whose gardens also extend to the river and they say that the Orators Isaac ffernandes Nunes Sr Philip Jackson John Becher & Moses Medina are inhabitants of some of these houses of a value of £800 upwards and the Orators William Shephard & Nathanel Shephard Infants are the Landlords & Owners of these houses and of eight others situated in the same Neighbourhood.

In his will[24] Medina left the Richmond house to his son-in-law. Jacob Zwarts records[25] a house named 'Richmond' on the left bank of the Vecht in Holland, which at one time belonged to the Capadae (Capadose?) family.

Medina, as noted above, lived in London until 1702, when he settled permanently in Holland.[26] He visited London occasionally afterwards, but always returned to Holland. He died on 4 Tishri 5491, *i.e.* Friday, 15 September 1730, when he was about 80 years old. His will was proved by Moses de Medina, his executor and son-in-law, on 1 October 1730. In accordance with his last wish, he was laid to rest in[27]

> Ouderkerk on the Amstel by the City of Amsterdam in the buryal Ground there belonging to the Portuguese Jewish Nation.

[20] P.R.O. C.11/1429/8.
[21] See p. 19.
[22] Medina's will (Appendix C).
[23] P.R.O. C.11/1429/8.
[24] Appendix C.
[25] *Nieuw Israelietisch Weekblad*, Amsterdam, November–December 1927.
[26] *Ibid.*
[27] *Cf.* his will. This is the famous burial ground in Amsterdam, opened in 1614, and continuously in use until World War II.

Ester, his wife, died some eight months later, on 14 Iyar 5491, *i.e.* Sunday, 20 May 1731, at a very advanced age. Probate was obtained by Solomon Hiskia de Medina on 9 June 1731, who became executor of her will on 1 April 1731 after his brother, Moses de Medina, the original executor, had died eleven days previously, on 20 March 1731. Ester de Medina was buried next to her husband.

I am obliged to the Portugees-Israelietische Gemeente at Amsterdam for letting me have photographs of the tombstones of both Solomon and his wife from the Ouderkerk burial ground. Unfortunately, the inscriptions cannot be reproduced here because they are almost invisible. According to the information of the Amsterdam Gemeente office, the stones were sunk far below the surface of the ground and it was necessary to dig down in order to decipher the text. They are made of freestone, and each is 2 metres long and 1.25 metres wide. (At that time the usual width of tombstones was 75 or 80 centimetres.) Their shape and inscriptions are as follows:

S^A	S^A
DA BEMAVENTURADO	DA BEMAVENTURADA
SELOMOH DE MEDINA	E VIRTUOZA ESTER DE
FO EM 4 DE TISRI	MEDINA DE ACEVEDO
DO ANNO 5491	FALECEO EM 14 DE
SUO ALMA GOZE DA	YIAR DO ANNO 5491
GLORIA	SUA ALMA GOZE DA
	GLORIA

CHAPTER TWO

Medina's position in the Jewish Community

Solomon de Medina occupied a high position in the Jewish society of his day, in England as well as in Holland. He interested himself in communal and cultural matters, thus symbolising the type of Sephardi Jew (but not confined only to Sephardim) whom the historian—because of the synthesis they personified of financial and communal activity—calls the 'species hollandia judaica'.[1]

In London he took a leading part in the affairs of the by now organised and rapidly growing Sephardi congregation. His name appears for the first time in the records of the Spanish and Portuguese Synagogue in London in 5430 (1670), when he paid his *Imposta* for the first time for the second half of that year.[2] Shortly afterwards,[3]

> On the 1st of Tisiri 5431 [15 September 1670], the Senhores of the Mahamad met in order to elect Hatan Torah and Hatan Berisit; and there came forth elected the gentlemen named below, which be to them *besimantob*:
> Sr Jacob de Miranda, *Hatan Torah*,
> Sr Selomoh de Medina, *Hatan Berisit*.

The honour bestowed upon Medina proves that he must have been held in high esteem by the Gentlemen of the Mahamad. For, in accordance with old Jewish tradition, only important men were chosen for the honour of Bridegroom of the Law or of Genesis.

---

[1] Sigmund Seeligmann, 'Die Juden in Holland: eine Charakteristik', in *Festskrift i Anledning af Professor David Simonsens*, Copenhagen, Herz's Bogtrykkeri, 1923, pp. 253-257.
[2] *El Libro*..., p. 47.
[3] *Ibid.*, p. 48. This is apparently a wrongly dated entry. According to the 9th *Ascama*, Bridegrooms were elected on the eve of Rosh Hashanah, which is the last day of the month of Elul (*El Libro*..., p. 6).

6

In Bevis Marks a special ceremonial accompanied the office of *Hatan* and was laid down in the *Ascamot*.[4] It began on the day after the election, *i.e.* on the first day of Rosh Hashanah, and lasted until after Shabbat Bereshith. During the reading of the Law in the Synagogue on Rosh Hashanah the Reader offered *Miseberah* for the *Hatanim* as an expression of the special honour bestowed upon them. But the main ceremonial was left to 'their festival', as it is termed in the *Ascama*, Simhat Torah and Shabbat Bereshith. One can well imagine that October day in 1670 when the Synagogue, decorated by the Bridegrooms with wreaths of myrtle, landscape tapestries, and gilt leather, as well as with flowers on their candelabra,[5] made a festive impression. The whole congregation was assembled when the moment arrived for both *Hatanim*, Miranda and Medina, to be accompanied by the Wardens, Isaac Israel de Francia and Abraham de Oliveira, to the special seats 'in their order as is customary.'[6] In accordance with the rights vested in *Hatanim* on 'their festival' during the service in the Synagogue, they distributed *Misvot*[7] to persons to be honoured, until they themselves were called to the reading of their respective portions from the Torah, Miranda first and Medina second.

The distinction awarded the comparatively young man Medina also proves that he must already have been a man of means, as it was either distinguished or wealthy men who were generally thus honoured, and the Treasurer's entries fully corroborate this. His *Imposta* for the last six months of 5430 amounted to £6 2s. od.[8]—the sixth highest for that period. Two and a half years later, in 5433, he paid £14 7s. 3d., being the highest *Imposta* paid for that half-year.[9] Medina himself must have regarded the Mahamad's decision to elect him *Hatan Bereshith* as a great honour, and he showed his appreciation by making valuable gifts to the Synagogue on this occasion. Thus we read in an Inventory, compiled in 5436/1676:[10]

No. 7. *Item*, two Sepharines that were purchased of the widow S^{ra} de Carvajal, for which the said S^{ra} received the value, one with a cloak of coloured velvet and the other with a cloak of coloured cloth, which

[4] *Ibid.*, p. 6 (9th *Ascama*).   [5] *Ibid.*, p. 74.   [6] *Ibid.*, p. 6.
[7] *Ibid.*   [8] *Ibid.*, p. 56.   [9] *Ibid.*, p. 70.   [10] *Ibid.*, p. 117.

[cloak] S^res Jacob de Miranda and Selomoh de Medina made Codez when they were Bridegrooms, and likewise another cloak of green cloth of the Synagogue.

The Inventory further contains the following entry:[11]

> Item, Hanuca-lights of copper which Sr Selomoh de Medina made Kodez.

It was unavoidable for a man of Medina's rapidly growing importance in business to play a parallel and continually increasing role in his congregation. Six years after receiving his first honour, he was already invited to join the Executive of the Synagogue in the capacity of Treasurer. The relevant entry reads as follows:[12]

> On the 12th Nisan, 5436 [26 March 1676] the Senores of the Mahamad met in order to hold the election of two Pernasin and one Gavay to serve this Kaal Kados for one year, and there came forth elected by ballot the gentlemen named below, which be to them *besimanthob*:
> Sr Abraham Rodriz de Francia, *First Pernaz*.
> Sr Isack de Paiva, *Second Pernaz*.
> Sr Selomoh de Medina, *Gavay*.

Four days after this election was the first day of Pessah and in accordance with the prevailing custom the new Mahamad was announced to the congregation.[13] The doors of the *Echal* were opened and the newly elected Executive members rose to promise 'to fulfil their duties with truth, justice, and fear of God, without respect or despite to the prejudice of parties, and this shall be observed inviolably'.[14]

Medina's duties from that date onwards were, apart from those vested in the Mahamad, specifically laid down for the Treasurer in the *Ascamot*.[15] He had to receive memoranda of *Impostas* from Yehidim, make up the accounts with the Parnas, and receive payments of *Impostas* and *Promesas*; it became his duty to record the decisions of the Mahamad in the *Book of Ascamot*; to read *Ascamot* from the *Teba* on intermediate Sabbaths on Succot and Pessah, and also to read from the *Teba* the accounts of *Sedaca* on Rosh Hashanah and enter them in the *Book of Ascamot*. He had to keep the proceeds of boxes and *Promesas* for *Terra Santa* and *Cautivos*. In case funds were needed by the *Hebra*, he had to advance these funds and was

[11] *Ibid.*, p. 118.   [12] *Ibid.*, p. 103.   [13] *Ibid.*, p. 104.
[14] *Ibid.*, p. 4.   [15] *Ibid.*, pp. 2, 7, 8, 13, 22, 23, 25, 50, 72, 96.

MEDINA'S POSITION IN THE JEWISH COMMUNITY 9

entitled to be reimbursed through the *Finta*. He also collected fines imposed for breach of rules of *Misvot* and for other reasons. At the end of the year he had to give to his successor an account with delivery of articles specified in the Inventory. A retiring *Gabay* was one of the electors of the new Mahamad.

In the first entry in the *Book of Ascamot* made by Medina as newly elected *Gabay*,[16] we read that the announcement of his and his two colleagues' election above referred to was made on 29 March 1676

> and S^r Ab. de Francia declined to accept his office.

He was the first *Parnas* in the Synagogue's history to decline, and accordingly a constitutional controversy arose,

> as it was a case that had not occurred in this Kaal Kados.

A special meeting was immediately called, in which Isack de Paiva and Medina, as the newly elected Executive members, as well as the former members of the Mahamad, Ishack Barzilay, Abraham de Oliveira, and Ishack Soarez Dortha, participated. They agreed to elect the former second *Parnas*, Abraham de Oliveira, as President. As a result of this unprecedented affair, Medina recorded in his accounts[17] the income of £10

> by the sum collected from S^r Ab. de Francia as fine for not accepting the office of Parnas.

This first entry by Medina in the *Book of Ascamot* is ironic, as twenty-five years later (5461/1701) he himself refused to accept election as *Parnas* and it was publicly announced from the *Teba* that he was fined £20.[18]

During his tenure of office as *Gabay* a number of organisational reforms were made in the London Kahal Kados. In the first place a fundamental alteration in the Synagogue's taxation

---

[16] *Ibid.*, p. 104.
[17] *Ibid.*, p. 111. The constitutional and legal implications of Francia's refusal to serve as *Parnas* are referred to in Neville Laski, *The Laws and Charities of the Spanish and Portuguese Jews' Congregation of London*, London, 1952, p. 19.
[18] *Minute Book*, pp. 37b, 41a; Moses Gaster, *History of the Ancient Synagogue of the Spanish and Portuguese Jews*. Memorial Volume to honour 200th Anniversary 1701–1901. 'Not published. For presentation only.' London, 1901, p. 53.

assessment was devised. No doubt Medina as Treasurer played an important part in this tax reform. Thanks to his ability in the field of finance and his experience in business he was well fitted to give valuable advice in this connection. According to an entry signed by him,[19]

> On the 14th of Elul of 5436 the Senhores of the Mahamad met, and considering the Ascamah which was made on the 10th of Tebeth of the year 5425 concerning the alteration of the ancient Imposta by common agreement of the heads of this Kaal Kados, and seeing the cause which necessitated the said increase to be now absent, they passed a resolution, and convened all the heads of families of this Kaal Kados in order to consult and agree upon what should be most fitting in this case;
>
> And they agreed unanimously to annul the Ascama which until this day has been followed for the payment of four shillings upon every £100 sterling, and they resolved that henceforward from the day of Ros Assana next of 5437 there should be due and paid two shillings on every hundred pounds of all goods and that should be bought or sold by their own or by others' hand on their own account or on commission in this city of London and its environs.
>
> *Item*, they likewise agreed that on all kinds of goods that pass in transit there should be due and paid one shilling on every hundred pounds;
>
> *Item*, that on all diamonds, rough or cut, or any other kind of gems that may be received to be despatched to another part there should be due and paid six pence on every hundred pounds; and the same shall apply to any kind of silver or gold that may come in or go out of this [city]; the said six pence shall be due and paid on every hundred pounds, and likewise on every hundred pounds that may be drawn or remitted in exchange to another part, and proportionately on smaller quantity;
>
> *Item*, likewise that on every hundred pounds placed on deposit account of their own or of others there shall be paid one shilling a year, and proportionately on smaller quantity; and the same shall apply to moneys put out on bottomry, on which shall be due and paid 1 shilling per £100.

Another important decision was taken during Medina's treasurership and recorded by him in the minutes. It refers to the education of children. While in general children were badly neglected in the London of Charles II, James II, and William III, and the charity school movement of religious societies to prevent children from vagrancy only commenced about 1696 or 1698,[20] the London Sephardim almost from the beginning

---

[19] *El Libro* . . . , pp. 105/106.

[20] M. Dorothy George, *London Life in the XVIIIth Century*, London, 1925, p. 220.

of their communal life made provision for the teaching of Hebrew (Bible and Talmud) to the adolescents and adults as well as for religious instruction for the children. This was among the many duties which the first Haham, Jacob Sasportas, had to fulfil when accepting the call to London in 5424 (1664).[21] As his duties—apart from teaching—were manifold, he was permitted to employ his son Samuel as a special assistant teacher.[22] The subsequent Haham, Joshua da Silva, when appointed in 5430 (1670), had to do the entire teaching by himself.[23] This may still have been possible then but the community grew steadily in numbers and it became increasingly difficult for the Haham to attend to all the functions which originally were bestowed upon him. It was during Medina's tenure of office that the appropriate reorganisation took place. The Haham was to be relieved from teaching children in the Talmud Torah and Isaac Israel Davila was appointed 'Reby' at a salary of £30 per annum,[24]

> to teach the Talmedim from Aleph Beth as far as reading Parasah in Hebrew (inasmuch as the other lessons from that point upwards devolve upon our Haham Sr Jehosuah da Silva).

A number of other measures and events are also recorded in the entries of Solomon de Medina, and are published in *El Libro de los Acuerdos* (pp. 104–113), with the exception of folio 44, which has disappeared from the original book.[25] All these entries conclude with the customary phrase 'and peace be upon all Israel', and are signed in his clear handwriting, of which a facsimile is here reproduced:

Medina's term of office as *Gabay* came to an end on 12 Nissan 5437 (14 April 1677).[26]

[21] *El Libro...*, pp. 15/16. [22] *Ibid.* [23] *Ibid.*, p. 41.
[24] *Ibid.*, p. 106. [25] *Ibid.*, p. 105. [26] *Ibid.*, p. 113.

There are in subsequent years a number of other decisions taken or events recorded in the Bevis Marks archives in which Medina participated in one way or another. Thus during the year 5437 it was felt by the leaders of the community that some *Ascamot* had to be modified in view of altered circumstances, and the first recasting of the Synagogue's constitution then took place. Among the signatories of the new version we find Medina's name.[27]

On 25 March 1700, as an Elder of the Congregation, he approved, together with others, the appointment of Daniel Peres as 'Sochete Bodech'.[28] He is recorded as an Elder for the last time in 5461 (1700/1701).[29]

Medina was not a signatory to the lease for the new Synagogue in 1699 but in 5460 he contributed £30 to the building fund.[30] The Synagogue was opened on 30 September 1701,[31] and six days later the very first wedding was solemnised there by Haham David Nieto, between Isaac de Joseph Yessurun Mendes and Ester de Abram Fernandes Nunes.[32] Solomon de Medina acted as one of the witnesses.[33]

Medina's seat in the new Synagogue was No. 3 in the second row on the right of the *Hechal*.[34]

The documents record that during the first six months of the existence of the new Synagogue Medina made various offerings totalling £9 16s. 0d.,[35] and in the last six months of the year —*i.e.* until September 1702—totalling £9 5s. 0d.[36] As in the previous year (5461), when he paid the highest *Imposta* for the last six months, £47 14s. 0d.,[37] he maintained this lead also in the last year of his stay in London, in 5462, with the record amount of £52 7s. 7d.[38]

The last act of Sir Solomon on behalf of the congregation in London was his signing of a lease on 30 July 1702, together with Alvaro de Fonseca, Isaac Telles da Costa, David Penso, and

---

[27] Reproduction of facsimile, *cf.* Gaster, *op. cit.*, pp. 12, 13.
[28] *Minute Book*, 5442, p. 36b.   [29] *Ibid.*, p. 38a.
[30] Gaster, *op. cit.*, p. 73.   [31] *Minute Book*, 5461.
[32] *Bevis Marks Records*, ed. Lionel D. Barnett, Part II, 'Abstracts of the Ketubot . . . from earliest times until 1837', 1949, p. 66, No. 87.
[33] Gaster, *op, cit.*, p. 104.   [34] *Minute Book*, 5462.
[35] Gaster, *op. cit.*, p. 91.   [36] *Ibid.*, p. 93.
[37] *Ibid.*, p. 88.   [38] *Ibid.*, p. 96.

Rowland Gideon, for tenements adjoining the Synagogue, for a period of 21 years at an annual rental of £40.[39]

From then onwards, his name does not appear again in the archives of Bevis Marks. He left England in the summer of 1702 and settled permanently in Holland.

[39] *Ibid.*, p. 120; *cf.* also Samuel, 'The First London Synagogue', *Trans. JHSE*, X, p. 127 (appendix II).

CHAPTER THREE

An Important Citizen

Important as was Medina's position within the Jewish community in England, it was not less remarkable among the non-Jews. Outstanding in this respect were his relations with King William III (William of Orange). These relations led some writers[1] to believe that Medina, like other Dutch Jews, had followed William to England in 1688. But it has been shown above[2] that Medina was living in London sixteen years before William's arrival in England.

William, like his predecessors of the House of Orange, when they became 'Stadholders' of Holland, treated the Jews in Holland exceptionally well. Members of the Jewish community were enabled to extend their trade freely, and under this Orange Prince they even had a chance of becoming consular representatives.[3] This not only benefited the Jews in the country but was also to the advantage of Holland herself, for it helped her considerably—thanks to the international connections of the Sephardi Jews—in attaining a position of great importance in the world's markets.[4]

It is not known how the new King and Medina became acquainted. But one can safely assume that it was through the Dutch Sephardi firm of Machado and Pereira that Medina's relations with the Government of England began and led to his personal contact with the King. This firm supplied William's

[1] See *Jewish Encyclopaedia*, VIII, p. 425; *The Universal Jewish Encyclopaedia*, 1942, VII, p. 437; Paul H. Emden, *Jews of Britain*, London, 1944, p. 25; Montague F. Modder, *The Jews in the Literature of England*, Philadelphia, 1944, p. 47; Lee M. Friedman, 'A Great Colonial Case and a Great Colonial Lawyer', in *PAJHS*, XLII (Sept. 1952), No. 1, p. 72.

[2] See p. 2.

[3] Simon M. Dubnow, *Weltgeschichte des judischen Volkes*, Berlin, Judischer Verlag, 1925-1929, VII, p. 464.

[4] Herbert Bloom, *The Economic Activities of the Jews in Amsterdam in the Seventeenth and Eighteenth Centuries*, Bayard Press, Williamsport, Pennsylvania, 1937, p. xv.

AN IMPORTANT CITIZEN

armies with bread and corn from the time when he became Stadholder in the Netherlands in 1672,[5] were his 'Providiteurs' during the campaign against Louis XIV in 1674,[6] and remained in this capacity throughout William's later campaigns.[7] In 1679 they supplied the English Army in Flanders with bread,[8] and subsequently became 'Providiteurs General of the English forces in the Netherlands'.[9] As a result of this it was a foregone conclusion that Machado and Pereira would be obliged to extend their business connections to England. The first time a London 'Assigne' of theirs is referred to in official English documents is in a letter from the Duke of Monmouth to Lemuell Kingdon, dated 7 August 1679.[10] But the Assigne's name is not given. There are no references at the Public Record Office in London to Machado and Pereira after 1681, but their name appears again after (Moses) Antonio Alvarez Machado took part in William's expedition to England in November 1688.[11] On 16 May 1690, Medina's name is mentioned for the first time in an official London document in connection with Machado and Pereira. It is in a communication from William Jephson to the Navy Commissioners of that date, instructing them[12]

> to issue (out of the £5,000 lent or to be lent to the Exchequer by Selomoh Medina on credit of the additional 12d. Aid) £5,000 to the Earl of Ranelagh to be paid over to Monsieur Machado and Pereira on account of Carriages furnished by them in Flanders.

[5] J. G. Van Dillen, 'De Economische Positie en Betekenis des Joden in de Republiek en in de Nederlandse Kolonial Wereld', in *Geschiedenis des Joden in Nederland*, ed. Hendrik Brugmans and Abraham Frank, Amsterdam, Van Holkema & Warendorf, 1940, pp. 561-616.
[6] Jacob Zwarts, 'De Joodsche Providiteurs General van den Konig Stadhouder Willem III,' in *Nederlandsche Israelitische Weekblad*, 18 July 1924.
[7] *Ibid.*
[8] *S.P.Dom.* Entry book S.P. 44/41, pp. 211, 246; 44/52, p. 141; 44/45, pp. 19, 25, 33.
[9] *Ibid.*, 44/45, p. 25; *S.P.Tr.* Index VI, p. 87. *Cf.* also Wolf, 'Postscript' to Bernard Shillman, 'The Jewish Cemetery at Ballybough in Dublin', *Trans. JHSE XI*, p. 165.
[10] *S.P.Dom.* Entry Book S.P. 44/51, p. 246.
[11] Nat. archief, Joh. de Jong 1803 (11 April 1724) in Zwarts, *op. cit.*; *cf.* also Van Dillen, *op. cit.*, p. 584.
[12] *S.P.Tr.* Disposition Book VIII, p. 135 (T.61/7-11).

On 21 May 1690[13] the King issued a warrant for the payment to Medina of £5,000

> as in full for carriages furnished the last summer by Monsieur Machado and Pereira to the English forces in Flanders.

Here for the first time the King instructed payment to be made to Medina. From that time onwards the files show a continual flow of petitions by Medina on behalf of Machado and Pereira to the King and/or the Lords of the Treasury and the instructions given by these authorities in accordance with his applications.[14]

As has been pointed out,[15] Medina's mother was a Pereira, and this fact may account for Sir Solomon's connection with Machado and Pereira. This connection was not in the capacity of employee or agent. In a Court deposition he is referred to as their 'Factor.'[16] Medina financed their contracts with the Government personally, or arranged finance on the basis of these contracts throughout the period they were acting as Government contractors. At the same time Medina gradually emerged in official circles in London in his own right and on his own merit as a man of importance and consequence. On 25 February 1690 he made a loan of £300 to the Government,[17] and increased it by £5,000 three months later.[18] His assistance was also requested by the authorities in highly confidential instances for the Secret Service. Thus in 1692 two bills of £100 each drawn from Mr. Guydett are referred to in the files in this connection.[19] On 23 January 1695 Medina's name is again registered[20] with £5,000 in a list of

> lenders of loans and amounts thereof upon security of the Government for prosecuting the war in Flanders.

He procured a further loan of £5,000 in 1696.[21] From then

[13] *S.P.Tr.* King's Warrant Book XV, p. 16 (T.52/15).
[14] *Cf.* for instance, *S.P.Dom.* S.P.8/10/No.117; *S.P.Tr.* T.1.28/82; T.1.37/7; T.1.46/59.   [15] See p. 1.   [16] P.R.O. C.9/300/8.
[17] *S.P.Tr.* Register of loans on the Additional 12d. Aid. Order Book III, T.60/3. Order No. 130.   [18] *Ibid.*
[19] *S.P.Tr.* Secret Service Money Account for the quarter from 25 December 1692 to 25 March 1693, T.38/735, ff. 43–51. This is of significance if read in connection with the reasons for the dismissal of Marlborough and Sir Solomon's testimony (see p. 65).
[20] *S.P.Tr.* Money Book XII, pp. 433–447. T.53/12.
[21] *S.P.Dom.* Vol. 86A. 1696, p. 320.

AN IMPORTANT CITIZEN

onwards Medina's loans to the Government became more frequent. They were mostly long-term loans but on many occasions he was approached for short-term advances to help to supply the bare necessities of the moment for the prosecution of the war.[22] These activities brought about written and oral negotiations between Medina and the Treasury officials. On some occasions, when Medina's petitions or contributions were dealt with by the Lords of the Treasury, the King was also present.[23] Thus the King became more and more acquainted with Medina's activities. It was as a result of these financial transactions that Medina was made one of the 'Commissioners for the Million Act Lottery' by Royal Warrant on 28 March 1694.[24] His duty was to become, together with others,[25] the

> Managers and Directors for preparing and delivering of tickets and to oversee the drawing of lots.

Over a year later, on 10 April 1695, Medina—like the other Commissioners—was paid £200 'as consideration for their services therein'.[26] There was only one other Jew who became a Commissioner for the Million Act Lottery at this time—Alvarez de Costa.[27]

Medina also had close contacts with the East India Company. The first reference to him appears in that company's *Court Book* under the date 29 July 1674,[28] registering his complaint against 'the narrowness of one bale of "Sovaguzes" bought at the last sale'. A similar claim is registered under his name on 28 August 1674,[29] and another on 23 February 1676.[30] On 18 September 1691[31] he was 'admitted to the freedom of the

---

[22] *Cf.* for instance, *S.P.Tr.* Treasury Minute Book VIII, p. 346 (T.29/8); IX, pp. 20, 106/107 (T.29/9); X, pp. 156–157 (T.29/10); XIII, pp. 195, 201, 205, 213 (T.61/13); XIV, p. 78 (T.1/14); p. 176 (T.53/14).

[23] *Cf.*, for instance, *S.P.Tr.* Treasury Minute Book VIII, p. 122 (T.29/8); IX, pp. 53, 58, 105 (T.29/9); XIII, p. 123.

[24] *S.P.Tr.* King's Warrant Book No. XVII, pp. 394–395 (T.52/17).

[25] *Ibid.*

[26] *S.P.Tr.* King's Warrant Book No. XVIII, p. 121, and Order Book IV, p. 172 (T.52/18 and T.60/4).

[27] *Ibid.*

[28] Ethel Bruce Sainsbury, *A Calender of the Court Minutes etc. of the East India Company* 1674–1676, Oxford, Clarendon Press, 1935, p. 64.

[29] *India Office* Court Book, Vol. XXIX, p. 35.

[30] *Ibid.*, p. 219.   [31] *Ibid.*, Vol. XXXVI, p. 73.

Company by redemption'. On 27 May 1698 he was appointed a member of the Committee to solicit Parliament for the preservation of the Company's privileges.[32] From 1694 until 1707 he is listed among the Adventurers in the East India Company.[33] From 1701 onwards his son-in-law, Moses de Medina, is also listed as Adventurer.[34]

While Medina did not participate in the original subscription of the Bank of England stock in June 1694, he is recorded as the first newcomer after the Books of the Bank had been opened.[35] On 25 August 1694, he held shares to the value of £1,000 and thus acquired voting rights, for which the minimum was 500 shares. In 1697 he became the third largest Jewish holder, with £3,110, and in 1701 the second largest Jewish stock-holder with £28,325. By 1709 his holdings were reduced to £16,600 and were finally disposed of in 1712. A fresh holding, amounting to £3,500, is recorded in 1722 and remained so in 1725. In 1727 it is down to £1,000 and as such recorded in Sir Solomon's will.[36]

In August 1697 Moses, Sir Solomon's son-in-law, acquired £750 of capital stock and voting rights.[37] In 1702, before his departure for Holland, Sir Solomon transferred to his daughter Deborah, Moses' wife, £200 capital stock.[38]

In his will Sir Solomon also apportioned £600 of Bank of England stock for his grandson, Solomon de Moses Medina, who had been a stock-holder since 1722.[39]

Sir Solomon never attended any meetings of the Bank of England and thus never made use of his voting rights. Moses, his son-in-law, attended once, in 1707. But, like many other Jews at the time, Sir Solomon frequently made use of the facilities accorded by the Bank.

Apart from the general business affairs which he conducted,

[32] *Ibid.*, Vol. XXXVII, p. 286.
[33] *India Office* Home Series, Misc. I–III.   [34] *Ibid.*, Vol. III.
[35] For this and the following references to the Bank of England entries I owe special thanks to Mr. J. A. Giuseppi, Archivist of the Bank of England, who permitted me to quote from his lecture, 'Sephardi Jews and the early years of the Bank of England', delivered in London before the J.H.S.E. on 18 March 1953. (*Trans. JHSE.* XIX, 1960, pp. 53–63).
[36] See Appendix C.   [37] Giuseppi, *loc. cit.*
[38] Sir Solomon's will (Appendix C).
[39] *Ibid.* and Giuseppi, *loc. cit.*

AN IMPORTANT CITIZEN

and to which reference is made in a separate chapter,[40] it is against the background of his activities referred to in this chapter that the first honour was accorded Medina: the King's visit to his house. The contemporary chronicler recorded on Saturday, 18 November 1699:[41]

> The same day his Majestie went to Hampton Court, where he will stay til Wednesday; and dined with Mr. Medina, a rich Jew, at Richmond.

We know from the *London Gazette*[42] that the King left for Hampton Court on 17 November. Accordingly it was on that day that the stately gilt coach drove over from the Palace to Richmond, stopped in front of Medina's house on the first rise of Richmond Hill, and William stepped out as the first King of England to pay a visit to a Jew. Unfortunately, no record of the purpose of this dinner is available. There may or may not have been a financial reason for the King's visit one day after his desperate appeal to the country through Parliament[43] for sorely needed money 'for the safety of the Kingdom by Sea and Land', for urgently needed repairs as a result of long wars, for discharging the 'Publick Debt', and especially for discharging debts from previous wars. Medina and his friends Machado and Pereira were particularly concerned in this last category, for there was £40,744 3s. 11d. still owing to them from the campaign in 1697 before the conclusion of the peace of Ryswick.[44] Did the King and Medina discuss a moratorium for the overdue repayment of this loan? This may well have been the case, and granted by Medina, because the debt referred to was only repaid after William's death in 1702.[45] On the other hand, at that particular juncture, new and additional funds were required and they may well have discussed a new loan. The documents keep silent on this. However that may be, the King's visit to the 'Jew Medina' in

---

[40] See pp. 23-25.
[41] Narcissus Luttrell, *A Brief Historical Relation of State Affairs from September 1678 to April 1714*. O.U.P., 1857, IV, p. 583.
[42] From Thursday, 16 November, to Monday, 20 November 1699.
[43] *Ibid.*, where a verbatim report of the King's speech of 16 November 1699 is given.
[44] *S.P.Tr.* XVII, pp. 1157-1160 (Army debts, T.38/798, pp. 73-75).
[45] *Ibid.*

November 1699 is in itself a unique event and worthy of particular emphasis.

Seven months later, on 23 June 1700, Solomon de Medina was knighted by William III at Hampton Court.

Peter Neve, Medina's contemporary, records this event in his *Pedigrees*,[46] in the chapter 'Knights and Batchelers', with the following words:

> Sr Solamon De Medina a Jewe Knighted at Hampton Court 23^d day of June 1700.

Le Neve, as he called himself, quoted as his source the *Partition Book*; he also listed Medina among the knights who paid the fee due by new knights for recording their knighthood in the Office of Arms. All my efforts to find Medina's coat of arms have failed. It is neither at the London College of Arms nor in any of the documents or printed literature which I consulted in the course of my research work. This strange absence of another clue to Sir Solomon's background deepens only what was termed elsewhere 'the mystery of Medina's origin'.[47] However, I obtained a photocopy of the page in the *Partition Book* (Vol. V, p. 59) with reference to Medina's knighthood, and reproduce it on page 21.[48]

It is not difficult to imagine that the non-Jewish population looked on this occurrence as an extraordinary one. The 'Jew Medina' became a Jewish symbol of honour and dignity. How important this honour appeared to the non-Jewish population can be deduced from the Decision of the Lords in Council and Session,[49] given in Edinburgh twelve years after Medina's

[46] Le Neve's *Pedigrees of the Knights (made by King Charles II, King James II, King William III and Queen Mary, King William alone, and Queen Anne)*. Ed. George W. Marshall, London, 1873, p. 473.

[47] See p. 1. Wolf includes Medina among the 'registered Sephardi arms' (*Trans. JHSE*, II, p. 159).

[48] The date in this document, January 1700, is old style; it is in fact 1701 new style. York here stands for York Herald. The signatures of the Heralds acknowledge receipt of the respective shares of the fee paid by Sir Solomon. *Cf.* also W. A. Shaw, *The Knights of England*, London, 1906, II, p. 272, who, without giving his source, recorded: '1700, June 23, Solomon de Medina, of London and Middlesex, a Jew (at Hampton Court).'

[49] *The Decisions of the Lords of Council and Session. From June 6, 1678, to July 30, 1712*. Collected by Sir John Lander of Fountainhall. Edinburgh, 1761, II, p. 708.

AN IMPORTANT CITIZEN

January 26th, 1700

Partition made in the College of Arms by Mr. Gromp York of One Knt. viz:

Solomon Medina (Jew) Knighted at Hampton Court 23. June, 1700.

			£ s d
Tho: St. George, G.r Sr. Thomas St. Geor.e Garter	— — — —		02 10 00
Hen: St. George Sr Henry St. Georg.e Clarenceux	— — — —		00 19 04
Sr John Dugdale Norroy	— — — —		00 19 04
Roft. Devenish — Robert Devenish York	— — — —		00 09 08
Henry Dethick Henry Dethick Richmond	— — — —		00 09 08
Gregory King Lancaster	— — — —		00 09 08
C. Mawson Charles Mawson Chester	— — — —		00 09 08
Peter Mauduit Windsor	— — — —		00 09 08
Samuel Stebbing Somerset	— — — —		00 09 08
Johan Gibbon Bluemantle	— — — —		00 09 08
Laur: Cromp Laurence Cromp Portcullis	— — — —		00 05 10
Newry L.C. Peter Le Neve Rougecroix	— — — —		00 05 10
Hugh Clopton Hugh Clopton Rougedragon	— — — —		00 05 10

knighthood. A Jew appeared before them as a witness, but was objected to by one party, who argued that Jews were 'inhabile in law, considering the rooted hatred they bear to all Christians.' But the Lords were of the opinion that as the case before them was not one between Jews and Christians, both parties being Christians, the Jew would not be prejudiced and should be admitted as witness. The Decision then continues:

> It was also remembered that the Queen[50] had knighted Sir Salomon de Medina, a Jew trading in London; and if capable of honours why not of bearing testimony?

[50] The Lords in Council overlooked the fact that Medina was not knighted by Queen Anne but by William III. Moses Dias in his Dedication also erroneously refers to the bestowing of the knighthood on Medina by Queen Anne (see Appendix E).

That day in June 1700 was indeed an important milestone in Anglo-Jewish history. Less than fifty years after the resettlement of Jews in England, Solomon de Medina was the first member of the Jewish community to be honoured in so striking a manner. Through him the whole community was honoured, for, as has been pointed out, he played a leading role in the affairs of his coreligionists in London. Medina thus begins the chain of the many knights of the Jewish faith to follow until this day. The fact that from that June in 1700 to the next promotion to knighthood of a Jew (Moses Montefiore), 137 years elapsed, in itself shows what an outstanding man Solomon de Medina must have been.

CHAPTER FOUR

Merchant

Medina's business activities were manifold. We know that on moving to England he brought along his 'goods from Amsterdam'.¹ But at that time he was in what we would term today the import and export business, like most of the Sephardi magnates. The endenization papers of Diego clearly show this, for they gave him the privilege²

> to pay noe more nor other Customes for any goods or Merchandizes by them imported or exported than his Maties naturall borne Subjects doe in like cases.

Medina's official title was 'merchant',³ and *The Little London Directory of 1677* records him as such.⁴ But no description more indicative is added. From the scarce information available we know for certain that as an Adventurer of the East India Company in London he purchased as far back as 1674 cotton goods like Sovaguzes,⁵ as well as Gelings.⁶ From a petition of his to the Lords Commissioners of the Treasury, dated 29 January 1695/6,⁷ we find that Medina was active in the coal trade. He purchased it in Newcastle, and thus, to my knowledge, became the first Jew in England to do so. As referred to in Appendix A, Medina is also listed among the merchants in silver.

Some further information as regards the nature of the merchandise he was concerned with can be gathered from the transactions of Sir Solomon's son-in-law, Moses de Medina, who some time after his marriage in 1692 began co-operation with his father-in-law. At first the newly married, eighteen-year-old Moses intended to establish himself in business as an

---

¹ *S.P.Dom.* Entry Book 36, p. 136; See Appendix A.
² British Museum Add. MS. 28074. See Appendix A.
³ *S.P.Dom.* Entry Book 36, p. 136.   ⁴ London, 1863, under (D).
⁵ East India Company, Court Book XXIX, pp. 35, 219.
⁶ *Ibid.*, XXXVI, p. 97.   ⁷ *S.P.Tr.* T.1.36/15.

23

independent merchant. But in 1693 a shipment of satin, calico, and 28 papers of Indian flowers arrived for him and was seized by the Customs.[8] Moses appealed against the seizure. On 6 July 1694, a Treasury Warrant to enter a *nolle prosequi* was issued.[9] It is said in the entry[10] that

> it appearing that said Medina is but a young beginner and a stranger to this sort of trade and he will no more enterprize any importations of this kind, wherefore the Queen is pleased to remit the Crown's share of the said forfeiture.

It is probable that Sir Solomon assisted his inexperienced son-in-law to extricate himself from this calamity, and afterwards enabled him to co-operate in his business. Apart from the merchandise referred to above, Moses' name is also connected with tin,[11] corals,[12] and gold.[13] Later, when the firm of Joseph Medina & Sons was established in Amsterdam,[14] Sir Solomon traded with them. This firm is known to have traded extensively in coffee and pepper[15] as well as in gold.[16] It can therefore be assumed that Solomon de Medina was trading in all those commodities mentioned.

Medina's turnover in the period before his entry into army contracts was considerable. We can judge this by inference only, but nevertheless to a high degree exactly. It has already been stated that Medina paid his *Imposta* to the London Sephardi Synagogue for the first time in 5430 (1670).[17] From then onwards he figures in the records of Bevis Marks as a regular contributor of *Impostas*. The computation of this 'income tax' was arrived at by a highly complicated method which, together with the rules governing this taxation, was laid down precisely in the *Ascamot* of the congregation.[18]

[8] *S.P.Tr.* Calendars, X, p. 659. [9] *Ibid.*, p. 696.
[10] It is of some interest that Moses' father, Joseph Aaron de Medina, is recorded as having committed a Customs offence a few years earlier (*S.P.Tr.* Calendars VIII, 1685–1689).
[11] *S.P.Tr.* Calendars, April 1705 to September 1706, p. 145.
[12] Correspondence, Memoranda (East India Company), 1725–1730, IX (unpaged).
[13] P.R.O. C.11/2731(159); also C.11/1727/4.
[14] See p. 26. [15] Bloom, *op. cit.*, Appendix C: IV, VIII.
[16] See p. 34. [17] See p. 6.
[18] *El Libro* . . ., pp. 1, 2, 4, 5, 18, 24–25, 30–31, 33, 37, 44, 46, 48–50, 54, 56, 58, 60, 62, 64, 69–71, 85–88, 97–99, 101, 105, 107, 111. (See also p. 10).

MERCHANT

If, on the basis of the calculation referred to above[19] and in view of the variety of goods as well as of the nature of the transactions, we accept the average rate of the *Imposta* as 2s. 6d. per £100, we can calculate approximately the extent of Medina's turnover in various years.[20]

Half year	Imposta	Turnover per annum
5430 (1670)	£6 2s. 0d.	£9,600
5431 (1671)	£2 0s. 0d.	£3,200
5433 (1673)	£14 7s. 3d.	£22,400
5434 (1673)	£20 0s. 0d.	£32,000
(1674)	£28 0s. 0d.	£44,800
5435 (1675)	£10 0s. 0d.	£16,000
5436 (1676)	£11 7s. 10d.	£18,000
5437 (1677)	£6 4s. 4d.	£9,700
5452 (1692)	£10 0s. 0d.	£16,000

The fluctuations in the *Imposta* figures may be due to the fact that the 'income tax return', if we may term it so, was lodged after payment for the goods was received, and owing to the slow communications and other technical difficulties, this was forthcoming only after considerable delays. Thus, for instance, an *Imposta* of £48 in 5434 may well have included transactions concluded in 5432, when no *Imposta* at all was paid. It seems to me that a fairer calculation—and nearer to the truth—would be to take the average of these years and estimate Medina's turnover in merchandise for the first eight years of his stay in London at about £20,000 per annum, which is a considerable figure for the time.

The colossal increase in his turnover in subsequent years can be measured by comparing the *Imposta* figures above with those of his last years of residence in London, when he paid £47 14s. 0d.[21] for the last six months of 5461 (1701) and £52 7s. 7d.[22] in 5462 (1702), thus reaching annual turnovers of £75,200 and £83,000 respectively.

But it was not as a merchant that Sir Solomon outshone his contemporaries. It was Medina the financier who left his mark on the economic history of his time, to which the next chapter will testify.

---

[19] See pp. 9–10.
[20] *El Libro* . . ., pp. 47, 56, 70, 86-87, 88, 98, 108. The figures for 5452 are taken from the Sedaca Account for 5452 (Bevis Marks Archive), p. 94.
[21] Gaster, *op. cit.*, p. 88.   [22] *Ibid.*, p. 96.

CHAPTER FIVE

Financier

(a) BILL DISCOUNTING, CREDITS, AND LOANS

There is no reference to any financial business in the early years of Medina's residence in London (1670–1689). As has been pointed out already,[1] it was through the Amsterdam firm of Machado and Pereira that Medina gradually switched over from merchandise to finance. That firm had a correspondent in London as far back as 1679, but the documents do not reveal whether Medina acted as their London representative then. His name is connected with them for the first time in 1690, shortly after William's coronation in England.[2] This refers to payments due to Medina for advances made by him to Machado and Pereira in the summer of 1689.[3] Wolf's assumption[4] that with the accession of William III Medina

> seems to have become the ornamental and devising head of a syndicate of army and war contractors, represented on change by the firms of Joseph Medina and Sons, of London, and Machado and Pereira, of Amsterdam,

is not warranted by the documents. The firm of Joseph Medina & Sons was established in Amsterdam (not in London) in 1702,[5] after the death of William III. Sir Solomon was never a member of this firm of his brother's and nephews',[6] but had from time to time 'considerable dealings in trade' with them.[7] Later, during his term of office as army contractor, he enabled them through Moses de Medina to participate in financing Government contracts.[8]

Nor was he—contrary to Wolf—head or partner of Machado and Pereira. A document was quoted above[9] in which Medina is called a 'factor' and this term means that he could 'buy and

[1] See p. 15.   [2] *Ibid.*
[3] *S.P.Tr.* King's Warrant Book XV, p. 16 (T.52/15).
[4] Lucien Wolf, 'Queen Anne's Army Contractors,' *Jewish Chronicle,* 28 June 1889, p. 16.
[5] See p. 14. P.R.O. C.11/1570/30.   [6] *Ibid.*
[7] *Ibid.*   [8] See p. 32.   [9] See p. 15.

sell in his own name, and had the possession and apparent ownership of the goods consigned, and a lien over them'.[10] He began as their financier and advanced certain funds to them against their bills. The first such loan of £5,000 against bills was granted by him in 1689.[11] This amount was used by Machado and Pereira in connection with their furnishing carriages (wagons) for the English army contingent which, as Marlborough's biographer states,[12] consisted then of eight thousand men, against the French in Flanders. The bills were repayable from funds due to Machado and Pereira by the King, and were to be made available after Marlborough's return to England.[13] But when the money was not then forthcoming, Medina submitted a petition to the King for payment on behalf of Machado and Pereira,[14] and finally received this loan back—by lending the same amount to the Government.[15]

As these complicated transactions constitute Medina's first direct financial dealings with Machado and Pereira against their own bills on the one hand, and with the Government on the other, the relevant documents should be quoted in full. Medina's petition reads as follows:

> May it please Your Ma^{ty}.
> Machado and Pereira having employed me Salomon de Medina, to Solicit the Payment of the £5000 which is due to them for the carriages they furnished, the last summer, in Flanders, for your Maties Service, and which, the Earle of Marlborough assured them, should be certainly paid, upon his arrival in England; I take leave humbly to begg Your Maj^{tie} to direct the speedy payment of that summe to me; they having had notice from the said Earle of Marlborough, that Your Matie had promised the payment of it, upon which they have drawn bills upon me.
> (signed) Salomon de Medina.

The Government obviously had no money, and Medina was approached to lend them £5,000 on Government security. He agreed, for an entry of 16 May 1690 reads:[16]

[10] *Chambers's Encyclopaedia*, ed. 1950, V, p. 562.
[11] *S.P.Dom.* S.P.8/10/No. 117.
[12] Sir Winston Churchill, *Marlborough. His Life and Times*, New York, Scribner, 1933–1938, I, p. 314.
[13] *S.P.Dom.* S.P.8/10/No. 117.
[14] *Ibid.*, and *S.P.Dom.* King William's Chest 10, No. 117.
[15] *S.P.Tr.* King's Warrant Book XV, p. 16 (T.52/15, p. 16).
[16] *S.P.Tr.* Disposition Book VIII, p. 135 (T.61/7-11, p. 135).

William Jephson to the Navy Commissioners, to issue (out of the £5000 lent or to be lent into the Exchequer by Solomon Medina on credits of the Additional 12d. Aid) £5000 to the Earl of Ranelagh to be paid over to Monsieurs Machado and Pereira on account of Carriages by them furnished in Flanders.

Medina paid over the £5,000 on the next day (17 May 1690),[17] and four days later[18] the King authorised payment to Medina:

Royal Warrant to Richard, Earl of Ranelagh, Paymaster of the Forces, to pay £5000 to Solomon de Medina, without deductions, as in full for carriages furnished the last summer by Monsieur Machado and Pereira to the English Forces in Flanders 'According to the usual rates allowed in those countries'.

From then onwards it became a usual feature for Medina to lend money to Machado and Pereira against their bills, and when the Government were unable to pay Machado and Pereira their dues, and the latter were thus unable to meet the bills, Medina lent money to the Government to enable them to meet their obligations. It was Medina's own money with which he finally repaid himself his loans, with the advantage of receiving from both Machado and Pereira and Government 'the usual rate'.

This 'usual rate' which he received from the Government was not fixed. It varied in accordance with the urgency of the funds needed, the amounts involved, and the time for repayment required. However, the basic rate of 6 per cent interest was always adhered to. On some occasions a further 2 per cent was granted[19] as 'an allowance (or premium or reward as additional to 6 per cent interest).' This can be contrasted with the Government's undertaking with the Bank of England, created in 1694, to borrow all the money subscribed (£1,200,000) at an interest rate of 8 per cent per annum.[20]

In most cases accounts were made out in Dutch guilders, and in this respect too, the rate of exchange was not fixed by the Government, but varied. On some occasions it was 9 guilders

[17] *S.P.Tr.* Register of loans on the Additional 12d. Aid. Order Book III. Order No. 130 (T.60/3).
[18] *S.P.Tr.* King's Warrant Book XV, p. 16 (T.52/15. p. 16).
[19] *S.P.Tr.* Money Book XIV, p. 176 (T.53/14).
[20] J. H. Clapham, *The Bank of England, a History*, 1944. See also P.R.O. C.11/2015/29 (No. 6).

FINANCIER 29

to the £ sterling,[21] on others 10 guilders per £ sterling,[22] on others 10 guilders 6 stivers[23] or even 10 guilders 17 stivers.[24]

A further benefit for Medina was derived from a discount granted to him on occasions of the issue of new or renewed securities for his loans[25] 'at the same rate as the King makes remittances in the like Species'.

The securities given to him by the Government were based on future incomes. Thus on some occasions he lent money on credit of the Additional 12d. Aid,[26] on others on the 3rd 4-shilling Aid,[27] or on incomes from New Tax,[28] on tallies,[29] on Exchequer Bills,[30] bills of exchange,[31] Bank bills,[32] or letters of credit.[33]

Medina's dealings with the Government gradually extended. Soon his loans were required not only to cover Machado and Pereira's debts to him but also for other purposes, which in the main were in connection with the conduct of William's wars. The first such direct loan, amounting to £5,000, was provided by Medina on 23 January 1694-5[34]

> upon security of the Government for prosecuting the war in Flanders. Ordered that these be repaid from the new Tax and entered on that Register.

On another occasion,[35] Medina attended a meeting held on 16 October 1696 at Treasury Chambers, Whitehall, in the presence of five Lords, and

> informs my Lords that he will draw bills for £8500 to be remitted to Amsterdam to take up as many of Mr. Hill's protested bills as will amount to that sum and my Lords promise he shall be repaid in 6 weeks time and in the meantime he shall have a deposit of £12,000 in tallies on

[21] *S.P.Tr.* Treasury Minute Book IX, p. 105 (T.29/9).
[22] *Ibid.*, p. 164 (T.29/9).   [23] *S.P.Tr.* Army Debt. T.38/798, p. 75.
[24] *S.P.Tr.* Treasury Minute Book XIII, p. 173.
[25] *S.P.Tr.* Treasury Minute Book IX, pp. 102/103, 105 (T.29/9).
[26] *S.P.Tr.* Disposition Book VIII, 135 (T.61/7-11), p. 135; Order Book III (T.60/3).
[27] *S.P.Tr.* Disposition Book XII, 221 (T.61/12, p. 221).
[28] *S.P.Tr.* Money Book XIII, pp. 433-447 (T.53/12).
[29] *S.P.Tr.* Treasury Minute Book VIII, p. 346 (T.29/8).
[30] *Ibid.*, IX, pp. 102-103, 105, 221 (T.29/9); XIV, p. 78.
[31] *Ibid.*, pp. 106-107, 33.   [32] *Ibid.*, p. 53.   [33] *S.P.Tr.* T.1./74/44.
[34] *S.P.Tr.* Money Book XII, 433-447 (T.53/12).
[35] *S.P.Tr.* Treasury Minute Book IX, p. 20 (T.29/9).

the Continued Impositions [part of those which were intended for the Duke of Savoy] to be his security; and if any loss accrue by this drawing or remitting, the King is to bear it, as long as Mr. Medina gives Mr. Abbot an Account thereof from time to time.

Similar transactions were performed throughout William's reign in England.[36] As far as can be ascertained, the conditions remained the same in regard to all these loans.

Apart from the loans granted to the Government, on which he received profits, Medina, as has been pointed out, also financed Machado and Pereira directly. As these transactions were of a private nature, no trace regarding the interest paid, or any other form of profit or remuneration, can be found in the official documents.

The twofold loans of Sir Solomon to the Government and to Machado and Pereira grew in volume as the wars continued. Thus, after six years of such activities, Medina attended in the presence of all the Lords of the Treasury a meeting held on 15 March 1697 at the Treasury Chambers, Whitehall, and informed them[37] that his arrears to date

> are now about £50,000 and a loss of £23,000 more on the tallies.

This was indeed a colossal amount, especially if one considers that in addition Medina continued to advance further amounts for the supply purposes of the armies. The Lords must have been amazed when Medina, claiming £73,000 arrears and losses, stated in almost the same breath that 'the service shall not stand still for this'. At the same meeting he agreed to enter into a new contract for bread and bread wagons in Flanders, and

> Mr. Medina says that he will take his payments in Exchequer Bills or otherwise at the same rate [of discount or exchange] as the King makes remittances in the like species.

The King, in full appreciation of Medina's magnanimous gesture, ordered at a meeting held in his presence and in that of all the Lords at the Treasury Chambers on 17 March 1697[38]

---

[36] See, for instance, *S.P.Tr.* Treasury Minute Book VIII, p. 346 (T.29/8); IX, pp. 53, 58, 106, 201 (T.29/9); X, p. 156 (T.29/10); XIII, pp. 195, 205 (T.61/13); XIV, pp. 33, 78 (T.61/14).
[37] *Ibid.*, IX, pp. 102/103 (T.29/9).   [38] *Ibid.*, p. 105 (T.29/9).

that the value of the Tallies when delivered to him be enquired into. As to Mr. Medina's arrear it's to be paid at the rate of 9 guilders to the £ sterling in tallies or the land Tax. As to his demand of loss by the last year's tallies my Lords are to make good as much as the discount of those Tallies amounted to at the time when delivered to him. And as to his advance on the Contract for this year he shall have Exchequer Bills at the discount and exchange rate at which money is remitted thereupon to Flanders.

This new contract involved new expenditure, but on the other hand some loans outstanding were repaid. This method was repeated time and again. The liabilities of one contract were not paid off while new contracts and new liabilities were entered into. At the time of William's death in 1702, the debts amounted to £40,744 3s. 11d.,[39] and, after certain adjustments, they were paid over to Medina and to Machado and Pereira respectively.[40]

According to their contract, Machado and Pereira were obliged to export corn from England for the English troops in Flanders. It was also Medina's arrangement with them to provide the corn. Whether this involved an additional profit or commission for Medina has not been ascertained. It seems that Medina did not, or could not, fulfil this particular part of the contract. Some disputes arose, until finally the King himself intervened, and on 27 December 1695[41]

> Directs my Lords to call Medina before them and if the covenant in his contract for exporting English corn be useless it must be left out; or it may be left in the contract though the performance shall not be required of him.

During that whole period it was one of Medina's further activities to arrange the shipping of 'the corn and other necessities for the army in Flanders.'[42] This was by no means an easy task, as an example will illustrate. An entry reads:[43]

> Treasury order to the Customs Commissioners to observe:
> Prefixing: (a) Order of the Queen in Council, dated Whitehall May 24th for permitting the ships, *Constant*, Samuel Morris, Master from London;

[39] *S.P.Tr.* Army Debt T.38/798, pp. 73–75.
[40] *Ibid.*; and *S.P.Tr.* Treasury Minute Book XIII, pp. 173–174; Out Letters XVII, p. 24 (T.27/17). See also pp. 15–16, 27–28.
[41] *S.P.Tr.* Treasury Minute Book VIII, p. 122 (T.29/8).
[42] *S.P.Tr.* Out Letters General, XIV, p. 385 (T.27/14).
[43] *S.P.Tr.* Out Letters Customs, XII, p. 407 (T.11/12).

Brother's Desire, Nicho. Beaker, Master from Colchester; *Industry*, Thomas Morris, Master from Woodbridge; *Elizabeth*, Nicholas Richardson, Master from Hull; *John*, John Harwood, Master from Hull; *Success*, Tho. Cockrell, Master from Woodbridge, to proceed on their Voyages to Rotterdam, Notwithstanding the embargo: All on the Petition of Solomon de Medina shewing that he has laden the six vessels with corn for the Army in Flanders, and the said corn will suffer damage by lying on board.

Medina not only paid for the corn, but also for the vessels. Having done so for a few years, he claimed the money back which he had laid out for the vessels. On 9 May 1695 he submitted a petition for that purpose, claiming that this 'ought to be paid for by their contract.'[44] It was only in connection with the dispatch of four ships a year later (March 1696) that this obligation was recognised in principle,[45] but as no immediate payment to Medina is traceable, it can be assumed that the advance payment for the vessels was added to his general loan account.

About a year before his departure from England Sir Solomon permitted his son-in-law, Moses de Medina, to co-operate with him in the finance of army contracts against bills. We know already that Moses had close business relations with Sir Solomon since the time he married the latter's daughter in 1692.[46] But it was only in 1701 that Moses was accepted into this important sector of Sir Solomon's activities. The first documentary evidence as regards this collaboration is dated 11 June 1701, and reads as follows:[47]

> May it Please Y^r Lord^{ps}
> Having Understood, that Y^r Lord^{ps} are treating about Ye Remittances for subsistance of the English Troops—Wee humbly Propose to doe it
> Att Ealeven Guilders and two stuivers Currant Money att too & three Usances—and—
> Att Ealeven Guilders att short sight or letters of Creditt provided that wee have the monies on the Delivering of the Bills
> (signed) Sal & Mosses de Medina.

As might be deduced from what has been said, Sir Solomon accepted Moses for his financial activities only with reluctance. This was partly because of Moses' lack of experience, to which

---

[44] *S.P.Tr.* Out Letters General, XIV, p. 385 (T.27/14).
[45] *S.P.Tr.* Out Letters Customs, XIII, p. 247 (T.11/13).
[46] See p. 23.   [47] *S.P.Tr.* T.1./74/44.

reference has already been made,[48] but also arose from the fact that Moses was very poor and a financial failure at an age at which Sir Solomon had already made a fortune. Recalling that period in 1712, Sir Solomon wrote with reference to Moses:[49]

> When I left England he was worth no more that what he gott in my house, by giving him a share in my Commissions, and what I gave him on my departure, with the Commissions of the Provedorie wch he owned was worth to him above £2,500, and was the foundation of his credit.

But Moses in the course of the years turned out to be a very capable and energetic agent for his father-in-law.

For approximately twelve years, Sir Solomon carried on in London his bill discounting, credit and loan activities until two events in world affairs brought a radical change and turning-point in his life: William's death and the War of the Spanish Succession.

(b) FINANCIAL SPECULATION

When William III died in 1702, and the War of the Spanish Succession broke out in the same year, Medina was over fifty years of age, and he decided to leave England and to take up permanent residence on the Continent.[50] As has already been shown, Medina had established close relations with Machado and Pereira by then, and it was only natural that he should continue to finance their supply of bread and bread wagons for the forces in the new conflagration. As in past campaigns, the official contractor for the years 1702–1706 was again Antonio Alvarez Machado.[51] And the contracting as well as financial business also was continued on the same lines as described above[52] for the period 1690–1702. Medina's role remained at first unaltered while he at the same time set out to prepare for considerably bigger activities.

Sir Solomon's removal from London coincides with the establishment in Amsterdam of the firm of Joseph Medina & Sons. Joseph Aaron de Medina was Sir Solomon's brother,[53] and he had three sons: Moses, Isaac (senior), and Solomon alias

[48] See p. 24.   [49] *S.P.Tr.* T.1./154/10A.   [50] See p. 13.
[51] *S.P.Tr.* Calendars, January 1703 to March 1705, pp. 12, 14, 18, 70, 75, 153, 170, 515, 536; *cf.* Medina's deposition of 6 December 1711, on pp. 49–50.
[52] See pp. 30–31.   [53] See Appendix B.

Francesco. Moses married Sir Solomon's daughter Deborah; Solomon alias Francesco married his niece Ester, who was Moses' daughter and Sir Solomon's grand-daughter; Isaac's daughter Rachel married her cousin, Solomon alias Diego, Moses' son, who was Sir Solomon's grandson. The family ties were thus closely knit. At the time of the establishment of the firm in 1702, Joseph de Medina lived in Amsterdam, as did Isaac senior,[54] while Solomon alias Francesco lived in London until 1714.[55] These last three mentioned[56]

> entered into co-partnership together and for severall years after carried on a joint trade and correspondence with divers persons for the benefit of the said co-partnership.

Moses de Medina[57] was not a partner in the firm at first, but he seems to have joined it at some later date, apparently after his father's death, for he is referred to as its London representative in 1728.[58] Sir Solomon[59] did not belong to Joseph Medina & Sons, but he 'had considerable dealings in trade' with them.[60] It is difficult to ascertain the nature of these dealings. We know that they were general merchants, and dealt in the import and export trade, among other things in coffee, pepper,[61] and gold.[62] They imported the gold into both England and Holland from various 'parts and places beyond the seas,'[63] but especially from Lisbon, where they had their own resident 'Correspondence and appointed factors... carrying on their said trade and merchandise.'[64] It is, therefore, quite possible that from the time of the establishment of Joseph Medina & Sons, they conducted Sir Solomon's merchandise business, because from that same year onwards (1702) he

[54] P.R.O. C.11.1570/30.
[55] Lansdowne 558, f. 69.
[56] P.R.O. C.11.1570/30.
[57] *Ibid.*
[58] *Ibid.*, Rachel de Medina's answer (17 November 1741) and Sir Solomon's will (Appendix C).
[59] P.R.O. Cll.1570/30.
[60] *Ibid.*, and P.R.O. C.11.1584/18 (both documents).
[61] Bloom, *op. cit.*, Appendix C: IV, VIII; also *The Decree in the Case of Solomon de Medina, Mosesson and Company, Merchants in London, and Roderigo Pacheco, Jacob de Lara and Manuel de Costa,* New York, 1728; and P.R.O. High Court Admiralty (H.C.A.) 42/10. See also p. 24.
[62] P.R.O. C.11.1727/4. *Cf.* p. 18. See about Joseph de Medina en Zonen in Van Dillen, *op. cit.*, pp. 584/585.
[63] P.R.O. C.11.1727/4.
[64] *Ibid.*

FINANCIER 35

himself devoted his time to new business ventures. In addition to this they participated through Moses de Medina in the army contracting business,[65] sharing with him in his arrangement with Sir Solomon one-third of the financial obligations of these contracts, as well as one-third of the profits. But in addition to their merchandising and bill-discounting relations, they were Sir Solomon's Amsterdam collaborators in his fast developing financial speculation. His London collaborator in this and in other respects was his son-in-law, Moses de Medina.[66]

Possessed of organising genius, Sir Solomon de Medina was predestined for the system of speculation which he devised and which was of daring conception. In his speculation the risk seemed—on the surface at any rate—immense; in fact, the attention he gave to the details of his work excluded any great risk, or, at least, reduced it to very small degrees. This was possible because he had first-hand information at his disposal and acted only on the basis of this information. Having established his agency in London, headed by Moses de Medina, and having arrived at a working agreement with Joseph Medina & Sons, of Amsterdam, he devised a system of expresses whereby he himself, at the rear of the English armies in their campaigns, which began in the summer of 1702, was able to inform his collaborators of the latest events from the battlefield —through special couriers—long before other speculators or sometimes even the Government were informed. In turn, his partners were buying and selling on the Exchange and on other markets in accordance with the news received. This was not the kind of speculation which at that time flourished among the rich Sephardim in Holland, who 'turned from commerce to speculation which seemed an easier and more genteel means of livelihood.[67] Though there is no doubt that Medina was fully aware of the speculation in stocks carried on in Amsterdam, which grew to such dimensions that 'women, children and greybeards were found dabbling in these companies',[68] with the

[65] See p. 26.
[66] In addition to his work for Sir Solomon and his dealings with Joseph Medina & Sons, Moses also acted as London correspondent for Jaques Henriques, of Amsterdam (P.R.O. C.24/1345, Town Depositions).
[67] Bloom, *op. cit.*, p. 186.
[68] M. F. J. Smith, *Tijd-Affaires in Effekten aan de Amsterdamsche Beurs*, Amsterdam, 1919, pp. 93ff., Bloom, *op. cit.*, p. 186.

catastrophic result well known from our own days, he was able to see the danger signs and tried to avoid them. In this he succeeded thanks to his organising genius. He was greatly helped by the very fact that he was continuously kept up to date with regard to the military position. For, as is described in the following chapter, it was part of the duties of an army contractor to keep sufficient numbers of persons in the field for the purpose of the swift movement of supplies, and it was even the duty of such persons to make their deliveries to the units at the front directly. For that purpose the contractor delegated to each brigade a clerk to whom the supply-distributing persons were subordinated. With regard to intelligence, their comings and goings to and from the field constituted the main channel of information. Thus Medina's work presupposed precise timing and great knowledge of the market, and the ability to correlate them. His success proves that Solomon de Medina mastered them all, and thus created a system which, as Cecil Roth[69] points out, was pursued by the Rothschilds a century later and established their fame. It is apparently on the basis of a paucity of information that Werner Sombart[70] calls Medina 'the founder of stock exchange speculation in England.' When Medina devised his system he no longer lived in England, and when stock trading began there in 1688,[71] Medina, although living in London then, had no part in it.

[69] *A History of the Jews in England*, Clarendon Press, Oxford, 1941 ed., p. 284n.
[70] *Die Juden und Das Wirtschaftsleben*, Munich and Leipzig, Duncker & Humblot, 1928, c. 1911, 1918, p. 106.
[71] Bloom, *op. cit.*, p. 186.

CHAPTER SIX

Army contractor

In the previous chapter Medina's arrangements as regards commercial and speculative activities are recorded. He made no less careful preparations with regard to the finance of army contracting. As pointed out already, the London pillar of his business structure became his son-in-law Moses de Medina. This applies to commerce and to speculation and to finance. The last was the most intricate of them all because it demanded the ability to negotiate with Government offices. Moses had a great teacher in his father-in-law and was thus able to swim smoothly through the heavy seas of Government contracts.

After Solomon's departure in 1702, almost throughout the period until 1711, Moses signed all the applications and petitions to the London authorities in Sir Solomon's name,[1] and—in the intervals of the latter's informal visits to England—conducted verbal negotiations and consultations with the authorities as the need arose.[2]

Moses thus gained an increasingly important personal position not only with the authorities but also in business circles in general. In the year when Sir Solomon left, Moses became an Adventurer in the East India Company,[3] and is recorded as such until 1707. Obviously, it was Sir Solomon's desire to leave a respectable representative behind in London, and he seems to have done everything in his power to build up his son-in-law's reputation.[4]

Until 1706, Solomon de Medina continued his collaboration with Antonio Alvarez Machado, who, as has been pointed out,[5]

[1] See, for instance, *S.P.Tr.* T.1.105/10; 106/31; 107/2; 113/47; 115/53; 119/31; 121/28, or Queens Warrant Book XXIII, p. 320.
[2] See, for instance, *S.P.Tr.* Out Letters (General), XIX, p. 44; T.1.114/62.
[3] India Office, Home Series. Misc. List of Adventurers in the East India Company, Vol. 3.
[4] His growing importance within the Jewish community of London is dealt with on pp. 18, 23, 24, 34, 35, 36.
[5] See p. 33.

was the main contractor for army bread and bread wagons. This collaboration was on the same basis as that conducted during Medina's activities in London, consisting in the main in negotiating the contracts with the Government in London and in financing the purchase of corn for the bread deliveries. Moses de Medina took the greater part of these duties off Sir Solomon's shoulders, for the latter's primary object at that period was the stock exchange speculation which he then developed. For that purpose his acting on behalf of Machado on the spot and in the area of Marlborough's armies offered the ideal background for his intelligence system.

In 1707, after Machado's death,[6] Sir Solomon became principal army contractor, a position he retained until 1711. It became his duty and responsibility to provide Marlborough's army with bread and bread wagons. His contracts, which were concluded separately for each campaign, were on the whole similar to those entered into by Machado and Pereira. It was by no means an easy task to take full responsibility for bread deliveries—upon which the army relied—as not only was the purchase of corn and the baking of bread required but also direct delivery to each brigade and regiment, which demanded a highly complicated organisational performance. To illustrate this, I quote two important clauses from one of Sir Solomon's contracts:[7]

> 4th Item—The said Sir Solomon de Medina does Covenant, promise, and grant to and with the said John Earle Powlet, Robert Harley, Esqre, Henry Paget Esqre, Sir Thos. Mansell Bar^t and Robt Benson Esqre their Executors Administrators and Assignes by these presents that he the said Sr Solomon de Medina his Heirs Executors Administrators and Assignes shall and will have in the field a Sufficient Number of persons of good Substance who may Answer and be obliged in Writing for the Performance of the Contents of this Contract and that for the greater Convenience of the forces the Bread shall from time to time be brought and delivered at the Head of each Brigade and of the Train of Artillery to the Respective Quarter Masters or such Persons as shall be thereto Appointed and to the End that the Distribution of Bread may be made the more regular and for the greater Ease of the forces the said Sir Solomon de Medina shall have a Clerke at the Head of Each Brigade, at the Train of Artillery and at the Head Quarters wth such other Person

[6] Zwarts, *op. cit.*, according to whom Machado died on 11 Tebet 5467 (16 December 1706, old style) and was buried at Ouderkerk.
[7] *S.P.Tr.* T.1.222/43 (Enclosure).

as shall be Requisite to assist him in the Delivery and Distribution of the said Bread the said Clerke shall take a Receipt or Acquittance in Writing from the Quarter Master or other Person of each Regiment or Train of Artillery thereto appointed under his Hand by wch Rec^ts are to be made upp and settled with such Officers every fortnight or three weeks at farthest the Contractors shall only furnish Bread to the Effective men and those that are present to Serjeants and Quarter Masters Inclusive and not to the Officers or their servants or Boys and to be assured of the true Accomplishment of this Article the Contractor shall not be liable to Deter Bread otherwise than on Lists of the men present and Effective of the respective Companys signed by the Major of the Regiment and in his absence by the Commanding Officers of the said Regiment who is to putt the same into the hands of the Generalls of the Horse and Foot respectively and in the garrisons to the Commanding Officers of the place who after Examination of the same shall give Orders for the Delivery of the Bread accordingly such Lists shall be putt into the hands of the Contractor from three to three Deliveries or every Twelve days upon pain that he shall receive nothing for what the Bread might cost more than the Soldiers are to pay and if by negligence of the Clerke any Dispute do afterwards Arise the loss shall fall on the said Sir Solomon de Medina his heirs Executors Administrators and Assignes if he cannot prove and make appease by the acquittances what they shall demand and as their due.

5th Item: It is agreed that if after such Acc^t made upp the Officers do not satisfye the said Sir Solomon de Medina his Agents or Assignes w^t they are to pay for the Bread they have rec^d upon his or their Representations and produceing to Pay Master the said Receipts and Accounts he shall forthwith pay and satisfy the same without any delay out of ye pay of ye said Officers Provided always that the respective Regiments or Train of Artillery be not oblidged to receive more Bread than they shall think necessary or shall have occasion for except by Order from the Gen^ll upon any March or other occasion In which case the soldiers shall be obliged to take the Bread provided for them or to Indempnify the s^d Sir Solomon de Medina his Executors Administrators and Assignes for the damage they shall receive thereby As her Majestye shall doe in the like manner for the Bread that may be bespoke and cannot possibly be brought to the Army and the Delivery of the Bread Expiring or Ending before the Middle Sepr 1711 the Contractor shall be Indempnifyed and relieved for the loss he shall happen to sustain of such corne or graine as he for the performance of the Delivery shall prove on due order to have laid by except the Gen^ll find good to lett the same be consumed by the forces on the same foot or condition as in this present contract.

It will be seen from these extracts that, apart from the organisational side of Medina's duties, clause 4 of the contract at the same time gives an indication of how he was able to organise his intelligence and receive almost daily first-hand news about the situation in the field, information which he was able to utilise for his financial speculations.

The contracts for supplying the armies were drafted so elaborately in order on the one hand to prevent any misunderstanding and on the other to safeguard the troops, who had to pay for the bread themselves in those days. The soldiers paid 5 stivers for a loaf of six pounds, while the contractor's price was 6 stivers 15 deniers, the difference being paid by the Government.[8] The Government auditor, E. Harley, defined the safeguards inherent in Medina's contracts in one of his reports as follows:[9]

> First, for restraining the Quantitys of the said Contractors Deliveries to the Numbers Present and Effective in each Regiment; and to check such Demands as he might make for the same, when the Price exceeded the Rate Paid by the soldier.
> And also for Preventing Disputes with the Troops about his Deliveries when the Value thereof came to be deducted or stopt from their Pay.

Apart from the payments referred to above, the contractor was paid a fixed amount for the wagons which he had to keep in the field. For 250 wagons he received 5 guilders 10 stivers per day.[10]

The detailed and highly elaborate organisational duties in the field were only part of Medina's obligations. The goods and carriages also had to be paid for, and the Government was very short of funds. Accordingly, Medina was forced to provide the finance. It was no longer the arrangement of additional funds, as in the days of Machado and Pereira, but from the moment of taking upon himself the duties of army contractor Medina had to arrange the full finance. However, this was beyond even his means. He continued to demand a payment on account from the Government when signing the contract, and this was complied with throughout the five years of his activities. The first contract for 1707 was concluded in February of that year, and on the eleventh of that month an advance payment of £24,651 3s. 3d. was made:[11]

[8] *S.P.Tr.* Reference Book VIII, p. 426.
[9] *S.P.Tr.* (Report) T.1.222/43.
[10] *S.P.Tr.* T.1.147/52 (Harley's Report).
[11] *S.P.Tr.* Out Letters (General) XVIII, p. 281; and Disposition Book XVIII, p. 250.

for the value of 100,000 Guilders as advance on the contract for bread and the like for 165,000 Guilders as advance on the contract for wagons whereof £19,720 18s. 7¼d. is to be charged on the order for the 40,000 men and £4,930 4s. 7¾d. on the order for the 10,000 men.

In February 1708 the contract for that year was signed and on 27 February the advance paid again amounted to £24,651 3s. 3d.[12]

> to answer 212,000 guilders, being four fifths of 265,000 guilders, agreed to be paid in advance to Sir Solomon de Medina, Contractor for bread and bread wagons for her Majesty's Forces in the Low Countries being for providing the same for the year 1708.

On 2 May of the following year, on signing his contract for 1709, Medina received an advance payment of £24,766 7s. 1d.[13] The documents indicate that in 1710 the advance was to be doubled, the Paymaster of the Forces being authorised to pay £50,325 15s. 0d. in advance.[14] But there were apparently not sufficient funds immediately available, and so Medina received only £20,000 on 31 March 1710[15] and the additional £15,325 15s. 0d. and £12,000 on 10 May 1710[16] respectively. The advance paid to Medina on 2 March 1711 on his contract for that year amounted to £10,000,[17] but in November an additional amount of £15,000 was made available to him,[18] so that he received £25,000 in advance in all.

These advances were not sufficient to fulfil all requirements, and, as has been pointed out, Medina had to finance the rest. He was not always able to do so himself, and therefore he made arrangements with individuals who provided finance and received a proportionate share of the profits. Thus, on his contract for 1707 Medina made such an arrangement with Leonard Vanderkaa,[19] a merchant of Amsterdam,[20] and a similar arrangement for 1709 with Jacob Hiskia Machado.[21]

[12] *S.P.Tr.* Queen's Warrant Book XXIII, p. 320; Out Letters (General) XVIII, p. 393; Disposition Book XIX, p. 111.
[13] *S.P.Tr.* Out Letters (General) XIX, p. 44; Disposition Book XIX, p. 284.
[14] *S.P.Tr.* Out Letters (General) XIX, p. 170.
[15] *S.P.Tr.* Disposition Book XX, p. 140.   [16] *Ibid.* pp. 159, 160.
[17] Treasury Minute Book XVIII, p. 167.   [18] *Ibid.* XIX, p. 94.
[19] *S.P.Tr.* Out Letters (General) XVIII, p. 281.
[20] P.R.O. C.11.2292/6.   [21] *S.P.Tr.* Out Letters (General) XIX, p. 44.

For the years 1710 and 1711 he combined with Adrian Vanderkaa,[22] an Amsterdam merchant, and Joshua Castano, an Antwerp merchant.[23] Side by side with these occasional partners, Moses de Medina was permanently in charge of the management for the years 1707–1710,[24] and his duty was to arrange for one-third of the money, for which he received one-third of the profits.[25] Moses shared his third with Joseph Medina & Sons, who helped him in providing his third of the finance.[26]

But even these financial arrangements were not sufficient to cover the expenses. Thus Medina was forced to reduce his investments in the East India Company to the trifling sum of £248 17s. 6d. in 1707,[27] having at one time reached the figure of £22,664 12s. 9d.[28] But in general he had to use the same method which Machado and Pereira were practising. He drew bills on a great number of individuals and firms, and soon found himself in trouble, because the Government were not in a position to meet their obligations towards him. This applies especially to the year 1708, because of the unusual length of the campaign. From then onwards the situation deteriorated more and more every year, because old debts were not paid by the Government—while new contracts were entered into. By 1712 Solomon de Medina was in such straitened circumstances that he was unable to continue as army contractor.

[22] P.R.O. C.11.2292/6.   [23] *Ibid.*
[24] *S.P.Tr.* T.1.154/10A.   [25] *Ibid.*   [26] *Ibid.*
[27] *India Office. Home Series. Misc.*, Vol. III.   [28] *Ibid.*, Vol. II.

CHAPTER SEVEN

Wealth and Decline

Medina brought his 'goods and merchandizes' with him when he arrived in England in 1672.[1] This shows that he was then already a man of means. The chapter above[2] which deals with his business activities confirms his financial standing from the outset of his stay in London. On average, Medina's turnover in his first eight years of residence in London amounted to some £20,000 per annum. It was because of his wealth (apart from his business ability and family relationships with them), as has been pointed out,[3] that Machado and Pereira availed themselves of his collaboration when they entered into army contracts with the Government. He was then already sufficiently strong financially to advance funds to Machado and Pereira, and at the same time to provide personal loans to the Government direct. By 1697 he claimed £73,000 arrears and losses due to him from the Machado and Pereira contracts, but he informed the Treasury that notwithstanding this debt he would continue to provide further funds for the Government.

It has been usually assumed that Medina made his fortune during the period of his contractorship for the supply of bread and bread wagons to Marlborough's armies. But there is no doubt—and this is proved in the preceding chapters—that when he became army contractor in 1707 he was extremely wealthy. From various documents we now know that on leaving the contractorship in 1712 he was on the verge of bankruptcy. Thus the commonly accepted assumption that he accumulated his wealth out of the army contracts is wholly incorrect and without any justification.

There is no doubt that Medina hoped to make handsome profits out of the army contracts, for otherwise he would not have staked colossal amounts on that venture, nor would he have committed himself to the various partners whom, as has

[1] See Appendix A. [2] See pp. 23–25. [3] See p. 27.

been pointed out,[4] he persuaded to co-operate in the financing of his contracts.

It was unfortunate for his calculations that Marlborough's campaigns in 1707/1708 lasted longer than originally anticipated. This caused very heavy additional expenditure on the part of the Government, which accordingly was unable to meet its financial obligations. On the other hand, the deliveries of bread could not stop, and Medina had to carry on.

The following schedule[5] gives a picture of Medina's additional expenditure arising out of the campaign in 1708:

Sir Solomon de Medina's bill of his extraordinary charges and losses by reason of the unusual length of the campaign in the year 1708 as submitted in French (but translated herewith) by M. de Medina to the Duke of Marlborough.

Memoire for the Prince and Duke of Marlborough.

	Florins:	stivers:	deniers:
for horses and wagons captured in the last campaign of 1707	2,080:	0:	0:
for the hire of 50 wagons from 1 Nov. 1708 to 2 Jan. 1709 (new style) at the rate of 5 fl: 10 st: 0 den: per wagon per day....	86,625:	0:	0:
1708			
13 June, 6000 loaves commanded by Monsieur Pendegrass and left at Ghent in the hands of the enemy: reduced to sacks of flour: 176			
13 July, by an order signed by the Comte de Bergeyk: sacks 860.			
22 August, by order of the said Earl: sacks 191.			
11 Sept. by a certificate of the carriers and of the Spanish Commissary of the sacks taken from the public magazines: sacks 3,828.			
30 Oct. by a certificate of Phelipi Cardosso, the Spanish Commissary: sacks 97, making a total of 5,152 sacks: which at 6 fl: 6 st: 8 den: per sack make	32,586:	8:	0:
12,000 fagots at 8 per 100	960:	0:	0:
8,000 empty sacks at 20 sols	8,000:	0:	0:

[4] See pp. 33–37, 41–42 [5] *S.P.Tr.* Queen's Warrant Book XXIV, p. 205.

WEALTH AND DECLINE

	Florins:	stivers:	deniers:
3 July: transport by order of General Cadogan of 4,000 sacks of flour from Malines to Brussels at 12 sols per sack .	2,400:	0:	0:
12 July: 3,950 loaves left at Louvain and sold at 2 sols each. Loss	759:	14:	0:
13 August: ordered by General Cadogan at Brussels and left there: 9,051 loaves spoiled and sold there at a sol each: Loss	1,527:	7:	2:
27 August: left at Ash 14,550 loaves which were not given out because of the sudden march and which became spoiled: Loss.	2,455:	6:	0:
9 Nov: paid to boatmen to Zeeland, Ostend, and back to Antwerp	2,199:	18:	4:
Wastage of flour during the said season, 116½ sacks out of 1,852 sacks . .	736:	17:	4:
Despatch of grain to Dixmuyden, Furnes and elsewhere: Loss of 12,000 empty sacks at 20 sols each	12,000:	0:	0:
1709, new style			
2 Jan: 62,000 to 63,000 loaves transported to Ghent for the Army, but by reason of the sudden march of Troops to their winter quarters, sold for one or two sols each as certified by a Notary, loss thereon	5,750:	0:	0:
3 Jan: 25,000 loaves left at Dendermonde but not given out by reason of the said march and sold as above: Total loss. . .	2,056:	15:	0:
loss of wagons and horses in the victualling train going to Brussels, likewise at Ostend, and in passing the Scheldt at Helle in front of the cannon.	20,000:	0:	0:
for the change of money drawn at Lille, Menin, and Courtrai (10 per cent on £152,320)	15,230:	0:	0:
for the augmentation of the grain after passing the Scheldt, at least 500 lost at rate of 35 gold gulden.	24,500:	0:	0:
	219,867:	5:	10:

The figures above do not show the total of Medina's expenditure in the campaign of 1707/1708, but only the extraordinary additions to the otherwise heavy bill. In March

1707 the Government already owed him an amount of £14,651 3s. 3d.,[6] and Medina seems to have encountered trouble from his creditors. Thus we read in a communication from Moses de Medina, on behalf of Solomon, to the Earl of Godolphin (then Lord High Treasurer), dated 23 March 1707:[7]

> I take leave humbly to crave Y^r Lo^{ps} Promise in the payment of the money remaining due to mee on the advance for Bread & Bread Waggons which amount to £14651: 3: 3, I having present pressing occasions for the same, having accepted a great quantity of bills of exchange which became due on mee daily, and must be punctually paid, and this sum being contracted to have been paid long since, and I never been so long postponed, I hope Y^r Lord^{sp} will be pleased to take into Your consideration to order me present relief it being almost impossible for mee at this time & jointure to subsist without this ready money, and it will be as impossible that the service beyond Sea be carried on not in due readiness. I therefore submit this urgent reasons to Y^r Lord^{ps} Great Justice & Wisdom, I remain with Great respect
>
> My Lord
> Y^r most humble & most obedient
> Servant
> M. de Medina.

In response, he received £5,000 on account,[8] but over a year later, in May 1708, Moses de Medina applied[9] to the Earl of Godolphin for the payment of the additional £9,651 3s. 3d., as Sir Solomon was apparently able to find some financial relief in the meantime. This is also confirmed by the fact that in his application for the payment of the £9,651 3s. 3d. Medina stated that[10]

> as I am sensible that ready money is not so plenty as it is to be wished, and I being very zealous for the Publick Service, and would avoid to trouble Y^r Lord^p any more, do humble beseech Y^r Lord^p to order me this day part of the said sum in money, and that the rest may be struck in Tallies on the grand Morgage in my name, and as this request is grounded in Reason & Justice, I hope Y^r Lord^{sp} will be pleased to cherish it, in directing mee the most Y^r Lord^{sp} can in ready money & tho I have a good quantity of said Tallies by me already, I am willing to do all what in me lies to forward the Publick Funds & will ever express with trewe respect
> I am, etc.

Here Medina expressed a similar attitude to that which he

[6] *S.P.Tr.* T.1.106/31.   [7] *Ibid.*   [8] *S.P.Tr.* T.1.106/31.
[9] *S.P.Tr.* T.1.107/2.   [10] *Ibid.*

had shown once before, when in 1697 the Government owed him £73,000 in arrears and losses, which notwithstanding he stated that 'the service shall not stand still for this.'[11] Thus Medina was willing to bear sacrifices for the country's interest, an attitude which shows the incorrectness and baselessness of the oft-expressed opinion that he only had material interests at heart.

It is quite obvious that Medina expected to recover his losses from new contracts, and this explains why, after the heavy losses sustained in 1707 and 1708, he entered into new agreements with the Government in 1709, 1710, and 1711. However, his expectations for recovery were not fulfilled, and he encountered misfortune after misfortune.

In March 1709 he entered into a contract for that year,[12] but by September 1709 it became clear to him that his calculations had been wrong. He could not have foreseen a great increase in the price of corn. The Government tried to help, and instructed James Brydges, the Paymaster of the Forces abroad,[13] to pay

> to Sir Solomon de Medina in advance on his contract for bread by reason of the extraordinary dearness of corn and to be made good to her Majesty again by the Deputy Paymaster abroad out of his deductions from time to time to be made for the bread delivered to the Forces ... £8,000 0s. 0d.

Soon after this an additional amount of £2,000 was paid to Medina under the same title.[14] But all this was not sufficient to cover the losses. The schedule for the years 1707-1709 printed above shows this clearly. The losses in 1709 alone amounted to £80,000.[15] Thus, instead, as he expected, of making good the losses sustained in 1707 and 1708, he had to add another colossal deficit for the year 1709. It may have been wise for him to stop there, but he apparently set his hopes on the next campaign. However, he only recognised his mistake three years later, in 1712, when he wrote to the Lord Treasurer[16] of

> that great and fatal losse I met with the year before [i.e., 1709] wch amounted, to four score thousand pounds sterling, and broke my Back.

[11] *S.P.Tr.* Treasury Minute Book IX, pp. 102-103 (T.29/9); see also this study, p. 30.
[12] *S.P.Tr.* Out Letters (General) XIX, p. 44.
[13] *S.P.Tr.* Disposition Book XX, pp. 45-46.   [14] *Ibid.*
[15] *S.P.Tr.* T.1.155/9.   [16] *Ibid.* See the letter in full below on p. 55.

In March 1710, through Moses de Medina, he entered into the contract for that year,[17] notwithstanding the fact that when accepting it at the meeting at the Treasury Moses demanded payment of the arrears for 1709,[18] and before and afterwards he wrote memorials to the Lord Treasurer in respect of these demands.[19] But the contract was entered into and the normal advance as referred to above[20] was duly paid to him.[21] However, the result was the same as in previous years. The loss of £80,000 for 1709 was increased by another £53,000 in 1710.[22] The situation became desperate, Sir Solomon came over to London, and together with his son-in-law attended a Treasury Board meeting on 21 March 1711.[23] It was the time of year when a new agreement was usually made for the ensuing year, and it seems that the Lords of the Treasury were again able to persuade Sir Solomon to enter into a new contract for 1711, while not paying him anything in the meantime in respect of his losses. The report on the meeting reads as follows:[24]

> Sir Solomon de Medina and his son[-in-law] are called in [Treasury Board meeting] and Sir Solomon's memorial is read:
> Send word to the Auditors of Impresto to be here on Friday afternoon at five o'clock about Mr. Brydge's accounts.
> Mr. Sloper is called in and upon reading a demand made by Sir Solomon de Medina of about £53,000 for bread and bread waggons and loss by the enemy the last year, my Lords observe that all the said demands (except what relates to the loss by the enemy and the extraordinary price allowed for the bread) that is the excess thereof, more than is to be deducted from the Troops remain provided for Sir Solomon either in the hands of Mr. Sweet or the respective agents for the several regiments of the said Troops and my Lords do direct that Mr. Brydges do write to Mr. Sweet by this night's post that their Lordships are very much surprised to hear that what ought to have been deducted from the Troops has not been paid over to Sir Solomon, and that his [Sweet's] answer thereto be returned as soon as possible.
> Sir Solomon's demands are referred to Mr. Brydges, who is to make a state thereof and to distinguish how much is provided for already by payments or remittances of bills of exchange sent to Mr. Sweet and how

---

[17] S.P.Tr. Minute Book XVII, p. 133; XVIII, p. 124.
[18] Ibid. pp. 45, 53.
[19] Ibid. pp. 92, 116–117, 139; S.P.Tr. T.1.113/47, 115/53, 119/31; S.P.Tr. Reference Book VIII, p. 426; S.P.Tr. Out Letters XIX, p. 244.
[20] See p. 41.   [21] S.P.Tr. Queen's Warrant Book XXIV, p. 205.
[22] S.P.Tr. Treasury Minute Book XVIII, pp. 183/184.
[23] Ibid.   [24] Ibid.

WEALTH AND DECLINE

much is to come out of the provision for extraordinaries in this Session of Parliament.

Medina entered into the new contract, apparently for the old reason that he might otherwise lose everything, but at the same time decided to take the negotiations regarding the payment of the arrears into his own hands. In that year— 1711—he spent some time in London, as he attended an occasional meeting at the Treasury and submitted memoranda on 29 June and 28 September.[25]

He then seems to have lost confidence in his son-in-law, as his own name now appears again in the documents, in contrast to previous years when Moses handled his father-in-law's affairs. We also notice for the first time that when Medina appeared before the Treasurers on 29 June 1711 he was not, as on former occasions, accompanied by Moses, but by Mr. Sloper.[26] From that date onwards no application or intervention on behalf of Sir Solomon is recorded in the name of Moses. In the meantime, Medina tried to get payments from the Government on the one hand and on the other from the partners to whom reference has been made above.[27] However, 1711 also turned out to be a year of deficit, and as this is the last year of his activities as contractor the accounts for that year are here set out:[28]

	Gilders
Sir Solomon de Medina is Debtor for ye money advanced ye 28th March 1711 on Acc^t of ye Wagons as Per Mr. Brydges Report	288420 – –
And for money advanced to enable him to provide Bread in ye beginning of ye year 1711 as per said Report	150000 – –
More Advanced on that accompt by Warrant of ye D: of Marlborough dated ye 26th of June 1711	50000 – –
In all 488420 – –	
Against which	
He appears by a Warrant of the Duke of Marlborough then Capⁿ Gen^{ll} of ye land Forces dated the 22nd December 1711 to be Entituled to ye sum of 469766 gilders for ye Service of 250 Waggons from ye 12 April 1711 to ye 15 May following inclusive and of 437 Wagons from ye said 15 May to ye 7th day of November following the said Warrant being yet unsatisfied.	469766 – –

[25] *S.P.Tr.* Treasury Minute Book XIX, pp. 25, 71. [26] *Ibid.*, p. 25.
[27] See pp. 33, 37, 41–42.   [28] *S.P.Tr.* T.1.147/52.

	Gilders
He also appears to have delivered to ye British Forces in Flanders during ye said Campaign 1711, Bread to ye value of 147160 Gilders 16 Sts 9^d as per Certificate from Cap^{tn} Henry Cartwright Mr. Brydges Deputy appears & which Captn Cartwright has orders from Mr. Brydges to stop from ye troops, and Sir Solomon affirms, that the proper Certificates are now in Mr. Cartwright's hand so remains due to Sir Solomon de Medina to discharge his whole demand on both ye said Contracts to ye 7th November last ye sum of	147160 16 9
	616926 16 9
	128506 16 9

It was at this moment, when Sir Solomon was desperately trying to save his neck, that the break with his son-in-law occurred, which must have been a heavy blow to him. (He was then in his 62nd year.) Their relations took such a turn for the worse that on 12 May 1712 Moses de Medina submitted a Memorial to the Lord Treasurer and Chancellor of the Exchequer 'complaining against Sir Solomon de Medina.'[29] Unfortunately this document is not traceable in the archives. From Sir Solomon's answer, however, the matter becomes clear, and as this affair caused Medina to break entirely with his son-in-law, it is worth reproducing in full.[30] In his Petition, an abridged version of which, quoted here, is recorded in another document,[31] he pointed out:

> that he gave his daughter in marriage to Moses de Medina with a very considerable fortune, that the said Moses pretends to be ruined in estate and reputation by means of £30,000 in bills drawn by him on the said Sir Solomon and on his account, they having come back protested; that the said Moses and his Father had combined together to let those Bills come back protested to ensnare him as acceptor, having before furnished him here with a good part of the money towards discharging of them which they deviated to their own private use and suffered those bills to come back protested; that his [Sir Solomon's] Accounts are now before the Auditors and that there appears to be due to him on balance about £25,000: therefore praying immediate payment of £13,000 to answer those protested bills.

The letter reproduced in full in Appendix D shows how bitter Sir Solomon must have felt about his son-in-law's appeal

[29] *S.P.Tr.* Treasury Minute Book XIX, p. 222.   [30] See Appendix D.
[31] *S.P.Tr.* Treasury Reference Book IX, p. 82.

WEALTH AND DECLINE 51

against him to the authorities. Moses de Medina may not have been aware of the steps taken by his father-in-law to recover the losses from the Government, nor may he have known that in his approach to the authorities Sir Solomon behaved very fairly indeed towards his son-in-law. A document constituting Sir Solomon's appeal for payment, and dated only three days before Moses submitted his petition against him, proves Sir Solomon's irreproachable stand in the matter. It reads:[32]

> To the Most Honble Robt Earle of Oxford & Mortimer
> Lord High Treasurer of Gt. Britain.
> Having had the honour last Tuesday solemnly to declare to Y^r Lord^sp that I could go no further with the paym^t of the Bills drawn & re-drawn for carrying on Her Maj^tys service, accepted for me by Moses de Medina whose credit although one of the best upon change lies at stake as well as mine, and we having both sold and pawned all we have in ye world, as Your Lord^sp may be easily informed.
> I take leave humbly to represent to Y^r Lord^sp that in case I am not presently succord with at least the sum of twenty thousand pounds our ruin and families must immediately follow, & in such a case Your Lord^sp can not doubt that the Credit of the Nation will certainly suffer, especially at this time when miscarriages both at home & abroad have been lately so fatal everybody well knowing the Governm^t is indebted to me about Sixty thousand pound, as shall appear by my Acc^t; all wch is most humbly submitted to Your Lord^sps great clemency & Equity by
> My Lord
> Y^r Lord^sps most humble most faithfull & distressed Servant
> May ye 9th 1712.   Sol. de Medina

Moses, as noted, probably did not know about this, and on 12 May 1712 he submitted his petition against his father-in-law and thus widened the breach between them which was to last for fifteen years, until 1727.[33] The documents, however, reveal that Moses' financial position was then by no means difficult. He confessed as witness in a lawsuit in 1727[34] that in 1708, 1709, and 1710 he dealt in stock of the Old East India Company and at the time of the Union in 1708 of the Old and New East India Stock he had 'upwards of £15,000' in his name, which he kept until 1725. This does not shed the most favourable light on Moses. When they became reconciled, Sir Solomon recognised—as he probably never denied—the legitimate claim of Moses, as well as of Joseph Medina & Sons, to the

[32] S.P.Tr. T.1.147/38.   [33] Medina's will (Appendix C).
[34] P.R.O. C.24/1443.

capital they advanced for the various contracts and for which they should have been credited with an equivalent profit participation. Medina settled with them in an agreement dated 20 May 1727, to which reference is made in his will, and according to which they were to receive 144,000 guilders (*i.e.* approximately £14,000).[35]

The difficulties which had arisen for Sir Solomon in 1712 were not confined to his partners, Moses de Medina and Joseph Medina & Sons. The great financial burden forced him to look round for other financial help as well, and we have seen[36] that he arranged a partnership for the campaigns in 1710 and 1711 with Adrian Vanderkaa, of Amsterdam, and Joshua Castano, of Antwerp. In view of his heavy losses, Medina was also unable to repay these loans as well as the interest or profits due, and therefore entered into an agreement with them on 27 May 1716[37]

> whereby it was agreed that all monyes which should be received from the Crown of England on account of their said claims should after such deductions as are therein mentioned be remitted to Mr. Abraham Muyssart then of Amsterdam Merchant since deceased for the Account of the said Sr Solomon de Medina Adrian Vandorkaa and Joshua Castano in the proportions following (viz) for the use and account of the said Sr Solomon de Medina $\frac{24}{40}$ or $\frac{3}{5}$ parts thereof and for the use and account of the said Adrian Vandorkaa $\frac{8}{40}$ or $\frac{1}{5}$ part and to the use of the said Joshua Castano the remaining $\frac{8}{40}$ or $\frac{1}{5}$ part. . . .

As a result of the pressure put on the London authorities by Medina, some arrears were paid in 1718 in Army Debentures, which were to be turned into South Sea Stock.[38] Accordingly, the three contracting parties entered into a new agreement on 15 September 1718[39] providing that the Army Debentures or South Sea Stock should be deposited with Abraham Muyssart and held by him on trust for the three partners in the same proportion as in the first agreement of 1716. After Castano's death a new agreement was made on 1 April 1727[40] confirming the old one and superseding the name of Castano by that of Bernard Van Tangeren representing the Castano family. When the partners regarded the time as appropriate for the

[35] *Ibid.* See also Van Dillen, *op. cit.*, pp. 584/585.
[36] See p. 42; P.R.O. T.1.145/4; C.11.2292/6.
[37] *Ibid.* [38] *Ibid.* [39] *Ibid.* [40] *Ibid.*

WEALTH AND DECLINE 53

sale of the South Sea Stock, they agreed to keep the proceeds in England and to authorise Hendrick Van Hatteren in London to hold the effects and funds on their behalf on a ½ per cent commission, while Gerard Venneck and Joshua Van Neck, merchants and partners in London, became their correspondents (the beneficiaries being Medina, Adrian Vanderkaa, and Van Tangeren) and had to take care of their joint concerns.[41] The effects were duly sold for £6,500, but Medina had died by then, and, while Vanderkaa and Van Tangeren received their appropriate shares, the Medina family did not receive their three-fifths part of the £6,500 nor any dividend thereof. Francesco de Medina, as executor of Sir Solomon's will, instituted court proceedings on 8 November 1731 for the recovery of the funds.[42]

Another partner in Medina's army contracts was Juda Pereira,[43] who apparently ceded his claim for payment to Machado in Amsterdam. Machado was unable to recover the money, and 'obtained a Decree in Holland against him for a very large sum of money.'[44] But during the submission to court of 'divers long accounts' in 1730 Medina died, and Machado thereupon 'obtained an attachment upon the Estate of Sir Solomon de Medina.'[45] The court proceedings lasted over ten years, and Sir Solomon's executor, Francesco alias Hiskia de Medina, settled the case on 9 June 1740 by paying 12,000 guilders (£1,200) to Machado.[46]

All these proceedings were the result of the losses which Medina sustained during his contractorship in respect of supplies for Marlborough's armies. The Duke of Marlborough, who as Commander-in-Chief had a thorough insight into affairs, testified[47] that from Medina's contracts with the Government Sir Solomon 'was much the greater loser by that.' Therefore, it is natural that when Medina saw in 1712 that he could not carry on any more, he 'declined contracting for the bread and bread waggons' for the next campaign, and informed the Treasury accordingly on 25 February 1712.[48] He thus gave up the idea of recovering his losses by

[41] *Ibid.*   [42] *Ibid.*   [43] P.R.O. T.1.145/4.
[44] P.R.O. C.11/806/11.   [45] *Ibid.*   [46] *Ibid.*
[47] *Marlborough Dispatches*, IV, p. 707.
[48] *S.P.Tr.* Treasury Out Letters (General) XX, p. 131.

accepting new contracts, as the old partners did not wish to continue and it would probably have been difficult for him to find new finance in the circumstances. The old debtors were pressing him hard, but at the same time they tried to get new contracts for themselves. At this point it should be mentioned that just at that time Medina's testimony in the Marlborough matter[49] contributed to the latter's dismissal, and some historians assume that Medina acted in agreement with the then Government. How unjustified such an assumption is—with which we deal below[50]—is proved by the fact just referred to that exactly four weeks after Marlborough's dismissal Medina was forced to decline a new contract offered him by the Government. This Government seems to have retreated behind the excuse of not having any money, and, without any consideration for the great services rendered by Sir Solomon, proceeded to appoint new contractors, without even trying to ease Medina's desperate financial situation. On the same day (25 February 1712), and in the same communication in which Medina's refusal to continue is recorded in the documents, Mr. Thomas Vernon, of Twickenham, was invited to call and discuss a contract.[51] A month later Solomon and Juda Pereira offered their services,[52] but finally Vanderkaa and Castano, Sir Solomon's former partners, were successful and received the contract.[53]

On leaving the post of army contractor Medina was a broken man. Bitter about the ingratitude of the Government, desperate about his position and about the demands made on him by his partners and creditors, he naturally tried to collect the arrears due to him. It took him more than eight years of unceasing effort. From among the many memorials submitted by him to the authorities between 1712 and 1720, I should like to reproduce only one, dated 4 December 1712, which reveals his innermost feelings as well as his desperate position:[54]

[49] See pp. 66–69.
[50] See pp. 66–70.
[51] *S.P.Tr.* Treasury Out Letters (General) XX, p. 131.
[52] *S.P.Tr.* Treasury Warrants XXI, p. 373. Juda Pereira is probably identical with the last partner of Sir Solomon of the same name (*vide supra*).
[53] *S.P.Tr.* T.1.145/4. See also p. 52.
[54] *S.P.Tr.*T.1.155/9.

WEALTH AND DECLINE

My Lord

Had it been my good fortune that Your Lordship were but rightly apprized of the hard circumstances, I am reduced to, by my zealous and faithful services to the Publick, I should hope (from your Lord^{sps} great goodness and justice) to be screened from the misfortunes, I daily labour under, it is therefore I humbly begg leave to acq^t Your Lord^{sp}, that in the year 1710 (after that great and fatal losse I met with the year before wch amounted, to four score thousand pounds, sterling and broke my Back) The sume of 240000 Guilders for paing the Bread furnished that year to the English Troops was left for me in Mr. Sweet's hand, w^{ch}, had I received it accordingly would have prevented all the misfortunes, that I since groan under, but by some mischance it was laid by, and no acco^{nt} had thereof till lately brought to light, by the provident care of the Honble Mr. Auditor Harley, for want of that money (in order to carry on the Service) I was forced to draw and redraw, many Bills of Exchange at a great disadvantage, which have been since Protested and contra Protested, and many of them, put in suit for, w^{ch}, all my Valuable goods in the Hague are taken in execution, and myself not safe to shew my face in London, Besides the vast Interest I have paid & daily pay, & must allow wth, I come to discharge the Bills (wch Ammounts to above £6000) about that sume My Lord is due to me on my Acct 1710, & after a far greater sume for the year 1711, as will be seen by my Acc^t, now lying before the Auditor near ready for Y^r Lord^{sps} perusal, & soe with all Submission I humbly pray Y^r Lord^p to order me the ready payment of the Ballance of that Acct, to prevent the ruyn that threatens him, who with the most profound Respect, Beggs leave to Subscribe himselfe,
My Lord
    Your Lordships most Humble & most
    devoted servant.
    S. de Medina.

Medina was partly successful, although he had to wait until 15 March 1713 for a payment of £17,760 12s. 11d. in respect of his losses in 1710.[55] As he lived in Holland, he was obliged to entrust a representative in London with the task of pursuing payment to him of the arrears. Owing to his break with Moses de Medina, he could no longer avail himself of the latter's assistance. Accordingly, in 1711, he appointed Captain Francis Stevens 'to Solicite and obtain for him the Severall Demands for Bread and Bread Waggons etc. due to him in the late Warr.'[56] Stevens acted on behalf of Sir Solomon until 1720,[57] and in that year the amount still due to Medina was £2,880.[58] Auditor E. Harley was ordered on 27 July 1719 to investigate

[55] *S.P.Tr.* Out Letters (General) XX, p. 357; Money Book XXII, p. 175.
[56] *S.P.Tr.* T.1.228/4.   [57] *Ibid.*   [58] *Ibid.*, and T.1.222/43.

this outstanding debt and found it correct.⁵⁹ While this was negotiated and finally agreed to, Captain Stevens applied to the Lords Commissioners of the Treasury on 18 May 1720⁶⁰ for the fees arising out of his work for Medina to be deducted from the payment of £2,880 as

> Sir Solomon now refused to pay your Petitionʳ what he contracted for and that Your Petitionʳ will be entirely ruined by the Debts he has contracted here in Soliciting the business of the said Sir Solomon unless speedily relieved.

Captain Stevens won his case, and Thomas More, the Deputy Paymaster, instructed 'Mr John Mendes da Costa at his house in St. Mary Acre London', who was to receive the money for Medina, to

> cause payment to be made att the Treasury for the fees due there and likewise that you will be so kind to adjust and pay Capᵗ Stevens his said Demands otherwise his Grace must be oblidged to deferr payment of £2880—0—0 till the Commissʳˢ for Determining the Debts of the Army be re-appointed which is shortly expected to whom then Captain Stevens must make his application for redress.

Thus ended the long and weary financial transactions of Medina, which in the end left him with but a small fraction of his once great fortune. His will proves this conclusively. The total bequests in it amount to some £20,000, and consist of the following items:

(a) *In Guilders*

(1) To the Talmud Thora [the Sephardi Synagogue] of the
Portuguese Jewish Nation *in Amsterdam*.   Guilders
   (a) on the first Sabbath after his death   25
   (b) to the Officers 'of the said Church' on the 7th day ...   50
   (c) for a particular legacy on the 7th day   500
   (d) on the 30th day after his death   50
   (e) eleven months after his death   50
(2) To the Honendal⁶¹ in *The Hague*
   (a) on the first Sabbath after his death   20
   (b) on the seventh day   30

---

⁵⁹ *Ibid.*   ⁶⁰ *S.P.Tr.* T.1.228/4.

⁶¹ 'Honen Dal' (*i.e.* Compassion for the Poor) was the name of the Portuguese Jewish Congregation in The Hague, organised in 1709 (D. M. Sluys, 'Hoogduits—Joods Amsterdam van 1635–1795', in Brugmans and Frank, *op. cit.*, pp. 309/310).

WEALTH AND DECLINE

		Guilders
(c)	for a particular legacy on the 7th day	500
(d)	on the 30th day	30
(e)	eleven months after his death	30
(3) To his wife		
(a)	'contents of her Ketuba'	15,000
(b)	an amount of	5,000
(4) To Joseph Medina & Sons		
(a)	on his death	72,500
	'by virtue and upon Account of the mutual Adjustment of Accounts and Agreement which already thereupon hath been passed under hand the 20th May last past 1728 and whereof a Notarial Act is yet to be executed'.	
(b)	'after the decease of both Testators a like sum also by virtue and in pursuance of the said Adjustment'	72,500
(5)	To Simhar Brandon, married to Aron Arias (London)	2,000
(6)	To Mrs. Ribka Aboab Ffonseca	1,900
(7)	To his clerk, David Roiz Monsanto	1,900
(8)	To the daughter of Abraham & Sarah Mementon	2,000
	'out of a particular Affection which the Testator hath had for the said Sarah Mementon which Sum the Testator hath promised by a written Act or missive as well as to her father Abraham Mementon as to her Husband David Usiel Cardoso'.	
(9)	To Sarah Mementon's brother Isaac	500
(10)	To 'his good friend Mr. Rabby Nunes Torres'	400
(11)	To 'his good friend Mr. Frans van Limburch' 2 Candlesticks.	
(12)	To his chamber servant if a Jew	250
	if of another religion	100
(13)	To his wife's chambermaid	50
(14) The Residue of his estate to		
(a)	his grand-daughter Ester de Medina	1,500
(b)	his grand-daughter Rachel Lamago	1,000
(c)	his grand-daughter Abigael Mendes	1,000
(d)	his grand-daughter Jochebeth 'in case she shall happen to marry with the consent and approbation of her parents'	3,000
	This brings the total to	181,885

	(b) *In Sterling*	Pounds
(15)	To the 'portuguese Church' at London called 'Sahar Asamaim' (Shaar haShamayim [this is the official transliteration])	50
(16)	To his wife (in South Sea annuities)	1,000
(17)	Into the executor's hands	400
	Altogether	1,450

It would be erroneous to assume that the amounts Medina bequeathed in his will were available when payments were due after his death. It has already been pointed out above that £1,200 due to Machado was only paid in 1740, and payment was made out of the estate, on which Machado had a charge. There were also other debts to be satisfied,[62] which remained unpaid for many years after Medina's death. Even the settlement of 72,500 guilders with Joseph de Medina & Sons, referred to in the will, was not fully paid off eleven years after Medina's death, 16,800 guilders (£1,620) being still outstanding then—in 1741.[63] This is the more symptomatic, as the final executor of Sir Solomon's will, Francesco alias Solomon Hiskia de Medina, was then also one of the last remaining partners of Joseph de Medina & Sons,[64] and would have allocated the funds to himself if these had been available. But there was 'not sufficient left to pay everybody.'[65]

As may be expected in a case where a wealthy relative passes away and bequests are not forthcoming, the descendants began to accuse each other of embezzlement, theft, and conspiracy, and in 1741 Medina v. Medina cases went to court. Sarah and Joseph, children of Abraham de Medina, 'Infants under the age of 21,' brought an action on 29 July 1741,[66] against Francesco alias Solomon Hiskia de Medina, as executor of Sir Solomon's will, for not paying them their dues as descendants, and accused Francesco of having mismanaged the estate and invested Sir Solomon's money in England and Holland to his own benefit, while at the same time refusing to give an account of the estate. Francesco, in his reply,[67] denied the accusations, stating that there were not sufficient funds available. He accused the first executor of Sir Solomon's will, Moses de Medina, of having, immediately after Sir Solomon's death, and in conspiracy with Sir Solomon's widow, Ester d'Azevedo, misappropriated the money. Francesco also pointed out that he handed Sir Solomon's accounts and books to 'Mr. John Brydges, Clerk in Court, to be inspected by claimants agents.'

Sarah and Joseph de Medina's reply is not preserved at the

[62] P.R.O. C.11.1570/30.   [63] Ibid.   [64] Ibid.
[65] P.R.O. C.11.807/2 (Francesco's answer).   [66] Ibid.
[67] Ibid. (18 September 1741).

Public Record Office. However, Francesco's answer to them[68] indicates their line of submission, which in the main was based on Francesco's unwillingness to present accounts. To this Francesco replied:

> The Defendant says that by virtue of this authority he got into his hands as much as he was able to of the 'several Estates and Effects' of the said Sir Solomon de Medina and of the said Lady Esther de Medina de Acevedo and has left with his clerk in court 36 pieces of paper, signed by himself which contain to the best of his knowledge a true account of all the respective real and personal estates of Sir Solomon de Medina and his wife respectively that came into his [the Defendant's] custody. And he says that these writings contain a true account of all Debts and Legacies as he has paid of the Testator's and Testatrix's money respectively, and all other sums that he has so paid on their respective Estates and when and to whom he has paid them.

He further denied the use of the funds for his personal gain, and explained that he distributed the funds which were available strictly in accordance with the respective wills.

On 10 September 1741 Francesco de Medina brought an action[69] against Rachel (the widow of Solomon alias Diego), Deborah (the widow of Moses and daughter of Sir Solomon), Abraham, Isaac senior, and Jochebet, for having arranged 'with the Governor and Company of the Bank of England, and with the Governor and Company of Merchants trading to the South Seas,' to divide among themselves £750 capital stock vested in Sir Solomon's name in the Capital Stock of the Governor and Company of the Bank of England, and £900 vested in the name of Ester de Medina d'Azevedo in Old South Seas Annuity Stock. Francesco claimed this money for himself as partner and under the title of the settlement with Joseph de Medina & Sons, as shown in Sir Solomon's will. Rachel de Medina denied the right of Francesco to these funds in her answer dated 17 November 1741,[70] and maintained that Sir Solomon's and his wife's will clearly allocated the funds to the descendants who now claimed them for themselves.

Similar answers were given on 17 June 1743 by Deborah de Medina[71] and Abraham de Medina.[72]

[68] P.R.O. C.11.806/11 (26 May 1742).
[70] *Ibid.* (Addendum).
[72] *Ibid.* (second document).
[69] P.R.O. C.11.1570/30.
[71] P.R.O. C.11.1584/18.

And so it went on. The quarrelling members of the family were apparently not aware of the fact that nothing near the amounts which they had all anticipated and hoped for was available in Sir Solomon's estate. All their claims, submissions, and answers were based on the assumption of some hidden fortune. But in the end the documents themselves, as submitted to the courts, revealed the true state of affairs, testifying as silent witnesses to the rise and decline of Sir Solomon de Medina's fabulous fortune.

CHAPTER EIGHT

Medina and Marlborough

It only remains now to refer to two events which made Sir Solomon de Medina known in his generation and commented on in successive ones: the affair of the Duke of Marlborough, and Voltaire's misfortune in London.

In view of the important role which Medina played in connection with Marlborough's dismissal, a few words must be devoted to their acquaintanceship and relations. Medina's activities in connection with Machado and Pereira, as well as his own contracts for supplying the English armies, soon brought him into contact with the great General.

After assuming full power in England, William III confirmed Marlborough in his rank of Lieutenant-General, and employed him practically as Commander-in-Chief to reconstitute the English Army. One month after the Coronation, in May 1689, war was declared against France, and, while William remained in England, Marlborough led the English contingent of eight thousand men against the army of Louis XIV in Flanders.

At that time, Machado and Pereira were the 'Providiteurs' of the English Army;[1] their London correspondent was Medina. In 1691 Marlborough's name crops up for the first time in a communication of Medina's. The latter submitted a petition to the King on behalf of Machado and Pereira, in which it is stated:[2]

> Machado and Pereira having employed me Salomon de Medina, to solicit the Payment of the £5,000 which is due to them for the carriages they furnished, the last summer, in Flanders, for Your Maties Service, and which, the Earle of Marlborough assured them should be certainly paid, upon his arrival in England; I take leave humbly to Begg Your Matie to direct the speedy payment of that sume to me; they having had notice from the said Earle of Marlborough that Your Matie had promised the payment of it, upon which they have drawn bills upon me.

[1] See p. 15.
[2] *S.P.Dom.* 8/10/No. 117 (of 1691–1692); Vol. 82, 1691–1692, p. 50.

Machado and Pereira also supplied bread to Marlborough's army in Ireland, and Isaac Pereira became 'Commissary-General of the Army in Ireland.'[3] Whether Medina also helped to finance these Irish contracts is not indicated in the documents.[4]

Medina and Marlborough became more closely associated during the War of the Spanish Succession. William III was dead, and his old Dutch 'Provideteurs' Machado and Pereira gradually fell into the background, while, with England's increasing importance in the war, Solomon de Medina's influence and scope of activities increased in proportion. In 1702 Medina accompanied the Commander-in-Chief in the campaigns on the European Continent and remained with him until 1711. For the last five years of this period, from 1707 onwards, Medina was the principal contractor to Marlborough's armies for the delivery of bread and bread wagons.[5]

It was natural for Medina's activities to bring him also personally nearer to the General. Marlborough's letters are about the only source from which we are able to get some indication of the personality and character of Medina, whom Marlborough called 'Le Chevalier Medina.'[6] On 31 March 1710 the General wrote to Moses de Medina:[7]

> [I] should be glad to oblige you in what you desire in behalf of your father;[8] but you must believe it will not be easy to persuade Sir Solomon to relinquish any part he has in the contract.

This shows that Marlborough regarded Medina as a strong man not easily amenable to compromise. The same letter ends with advice from the General again confirming this characteristic:

> I must now repeat what I told you before, that though you should think you had never so much reason to complain of any hardship from Sir Salomon, you must take care not to let him perceive you are under any manner of uneasiness with him.

Such understanding of Medina's personality presupposes a thorough knowledge of the man on the part of Marlborough,

[3] *S.P.Dom.* Warrant Book 35, p. 200; also (80) Vol. 1689–1690, p. 514.
[4] Wolf assumes that he did so ('Postscript', *op. cit.*), p. 165. [5] See p. 38.
[6] *Marlborough Dispatches*, III, p. 322.   [7] *Ibid.*, Vol. IV, p. 707.
[8] *I.e.* father-in-law.

which again is an indirect proof of their close acquaintanceship.

When Queen Anne dismissed Godolphin on 7 August 1711 and a new Cabinet was formed, an investigation was initiated into the financial conduct of their predecessors. The investigation was instigated in the last resort to force Marlborough to agree to a separate peace with France, or else to take upon himself the odium of having his private affairs dragged into the open. Marlborough insisted on the continuation of the war; the investigations took their course.

Owing to the close collaboration of Medina with Marlborough, it was a foregone conclusion that he, too, would have to appear before the investigating commission and testify about his financial relations with the General. Medina, who then lived in The Hague, was summoned to London at the end of 1711. When this became known to Marlborough, he wrote to Medina[9] 'that he was glad he was to be a witness and would afford any documentary assistance in his power'.

On 6 December 1711 Medina appeared before the commission of investigation and made the following deposition:[10]

SIR SOLOMON DE MEDINA Knight, being sworn on the Pentateuch deposeth that from the year 1707 to this present year 1711 both inclusive he has been solely or in partnership concerned in the contracts for bread & bread waggons for supplying the forces in the Low Countries in the Queen of Great Britain's Pay and that he gave his Grace the Duke of Marlborough for his own use the several sums following viz. for the year 1707 Guilders 66,600. 1708 G62,625. 1709 G69,570.15s. 1710 G.66,810.19.8. TOTAL Guilders 265,614.14.8 and also G.21,500 for this present year in part of a like sum with those above mentioned all which sums he gave his Grace because the former Contractors had given the like annual sums. He further deposeth that he allowed yearly 22 Waggons Gratis to the General Officers 12 or 14 of which were for the Duke of Marlborough's own use and that the former contractor did the same. This deponent further saith that from the said year 1707 to this year 1711 both inclusive, he gave yearly on sealing the said contracts a gratuity of 500 gold ducats to MR. CARDONNEL Secretary to the Duke of Marlborough for his trouble and pains in translating the Dutch Contracts and putting the English Contracts into form. And he further saith that for all the money he received in Holland from MR. SWEET deputy Paymaster at AMSTERDAM on account of the said contracts he was obliged to pay him 1% for prompt payment, that the former contractor did the same but he found him notwithstanding so backward in his payments that he complained to the Duke of Marlborough and at the

[9] Churchill, *op. cit.*, IV, p. 483.   [10] British Museum, Add.MS. 33273. F.151.

same time acquainted him with the allowance he made MR. SWEET of 1% as aforesaid and that his Grace reproved him for not paying this deponent more punctually and this deponent further saith that it appeared by the accounts of ANTONIO ALVAREZ MACHADO who had supplied the bread & bread waggons for the forces in the English pay as aforesaid for the years 1702, 1703, 1704, 1705, 1706, that he gave as large yearly sums to the Duke of Marlborough as this deponent hath done since.
Jurat 6th December, 1711.

On 15 December the Commons passed an Order for the commissioners of public accounts to report their proceedings.[11] In accordance therewith, on 21 December 1711 the House was presented by Mr. Lockhart, one of the leading commissioners, with the *Report of the Commissioners for Examining and Stating the Publick Accounts—the Duke of Marlborough's Accounts—And the Affairs of the Army*.[12] The passage in this Report dealing with Medina's testimony enlarges in some respects the original deposition, and is therefore worthy of recording here in full:

> Your Commissioners having ground to believe, that there had been some Mismanagement in making the contracts for the use of the Army, summoned, and examined, Sir Solomon de Medina, the Contractor for the Bread, and Bread-Waggons, in the Low countries, who, after expressing much uneasiness on the apprehensions he had of being thought an informer and of accusing a great man, did depose: That for the years 1707, 1708, 1709, 1710, and 1711, he has been solely, or in partnership, concerned in the contracts for supplying bread, and bread-waggons to the forces in the Low countries in the Queen of Great Britain's pay; and that he gave to the Duke of Marlborough, for his own use, on each contract, the several sums following; part of which was paid at the beginning and part at the end, of each respective contract, in bills or notes delivered by the deponent into the duke's own hands; viz. for the year
>
> | | |
> |---|---|
> | 1707................ | 66,600 guilders |
> | 1708................ | 62,625 guilders |
> | 1709................ | 69,578 guilders, 15 stivers |
> | 1710................ | 66,810 guilders, 14 stivers, 8 pennings |
> | 1711................ | 21,000 guilders first part for the year |
> | | 286,614 guilders |
> | | 45,810 guilders intended to be paid |
> | | 332,425 guilders, 14 stivers. |

[11] William Coxe, *Memoirs of John Duke of Marlborough, with his Original Correspondence*, 6 v., 2nd ed., London, Longman, Hurst, Rees, Orme & Brown, 1820, VI, p. 149.

[12] *Cobbett's Parliamentary History of England from the Earliest Period to the*

From whence it appears, that the Duke of Marlborough has received on account of the bread, and bread-waggons contracts, from Sir Salomon de Medina (admitting the sum already paid, and that is intended to be paid for this present year 1711, to be the same with that of the preceding year, 1710) 332,425 guilders and 14 stivers. From Antonio Alvarez Machado, like sums, which together make 664,851 guilders 8 stivers, and computed at 10 guilders 10 stivers to the pound sterling amount to £63,410 3s. 7d.

The commissioners who investigated the case and examined Sir Solomon de Medina were Geo. Lockhart, Hen. Bertie, S. Winnington, Fra. Annesley, Theo Lister, Will. Shippen, H. Campion.[13]

Thanks to a whispering campaign artificially created to discredit Marlborough, Medina's deposition became known before it was submitted to Parliament,[14] and this explains why, on 10 November 1711, the Duke of Marlborough wrote a full explanation to the Commissioners from The Hague, where he then was. Therein he said:[15]

> Gentlemen:—Having been informed on my arrival here that Sir Salomon de Medina has acquainted you with my having received several sums of money from him, that it might make the less impression on you I would lose no time in letting you know that this is no more than what has been allowed as a perquisite to the general, or commander-in-chief of the army in the Low countries, even before the Revolution, and since; and I do assure you, at the same time that whatever sums I have received on that account have been constantly employed for the service of the public, in keeping secret correspondence, and getting intelligence of the enemy's motions and designs.

The Commissioners, in summing up Sir Solomon's deposition and the Duke's answer, came to the conclusion that:[16]

year 1803, London, Hansard, 1810, IV, pp. 1049/1050; also for the following quotations.
[13] *Cobbett's Parliamentary History*, VI, p. 1056.
[14] Coxe, *op. cit.*, VI, p. 124.
[15] The letter appears in full in *Cobbett's Parliamentary History*, VI, pp. 1050/1051. I reproduce the relevant part of the letter only. It further deals with a number of financial matters in detail, but these do not relate to Medina and are, therefore, omitted.
[16] *Ibid.* p. 1052. Marlborough also prepared a statement for the sitting of Parliament on 24 January, which entered into still more detailed reply to the challenge. (*Cf. Cobbett's Parliamentary History*, IV, pp. 1079/1088.)

> as to what therefore relates to the evidence of Sir Salomon de Medina, his grace has been pleased to admit it in general, but with this distinction, that he claims the sums received, as perquisites to the general in the Low countries. On which your Commissioners observe, that so far as they have hitherto been capable of informing themselves in the constitution of the army, the great sums . . . can never be esteemed legal or warrantable perquisites.

Parliament then adjourned on 21 December 1711 until 17 January 1712, and an Order was passed by the House requiring the Commissioners to produce the documents on which their statement was founded.

But during the recess, on 30 December 1711, the Queen dismissed Marlborough 'from all his employment, that the matter might take an impartial examination'.[17] When the Commissioners' Report came before Parliament, on 24 January 1712, a majority of 265 to 155 decided 'that the taking several sums of money, annually by the Duke of Marlborough from the Contractors for furnishing the Bread and Bread-Waggons in the Low-countries, was unwarrantable and illegal'.[18]

Thus it was not ultimately on Medina's testimony that Marlborough was disgraced, although it is obvious that the whole investigation was only a manœuvre to discredit the General, in order to open the path to a separate peace with France. Medina himself, as he said, felt 'much uneasiness' in testifying in the matter, but, called to testify, nothing was left to him but to tell the truth. No small sensation was caused by the knowledge that the great General had been defeated by his enemies with the help of a Jew's testimony. In a passionate article on the subject, a writer in the *Examiner* said:[19]

> Neither can I think that he will be much disgusted though not One of the inferior Clergy should, in spite to *Sir Solomon*, make him a greater Man than *Moses*, nor any Mitred Orator extol his Name in a Place set apart for the Praises of his Maker.

And in the same paper an epigrammatist satirically immortalised this episode in the verse:[20]

[17] *Ibid.*, p. 1058.   [18] *Ibid.*, p. 1077.
[19] 6 November to 13 November 1712. Some writers, for instance, F. Modder, *op. cit.*, attribute this article to Jonathan Swift, but the latter contributed to *The Examiner* from 2 November 1710 to 14 June 1711 only.
[20] *Examiner*, 14 April, 1712.

A Jew and a G–n–l both joined a Trade,
The Jew was a Baker, the G–n–l sold bread.

Jonathan Swift, their contemporary, wrote that[21]

> Sir Solomon Medina paid him [Marlborough] six thousand pounds a year to have the employment of providing bread for the army.

Many subsequent writers and historians repeated this erroneous statement. Disraeli wrote of Marlborough and Medina as partners.[22] Even Wolf accepted this statement uncritically when he wrote:[23]

> His [Medina's] success in gaining these contracts was due to the personal friendship and interest of Marlborough, which he purchased by an annual 'present' to that commander of £6,000.

No wonder that antisemitic writers made full use of such an opportunity of proving 'Jewish character and Jewish custom as the cause of such irregularities' and of 'acquiring contracts through shameless bribery'. One of them[24] devoted a long chapter to this titbit. This shows where an erroneous statement made by a responsible person can lead. For in this case, fortunately, the documents bear testimony that Marlborough never arranged any employment for Medina, and never negotiated or concluded any contract with him. As we have seen,[25] the first agreement was made in February 1707, when the contract for that year was negotiated with the London authorities by Moses de Medina on Sir Solomon's behalf. It was concluded as the result of direct negotiations between contractor and Government. All the successive contracts were, throughout, also negotiated and concluded direct with the Government.[26] No intervention on the part of Marlborough, no support for the agreement nor any other use of influence by the General on Medina's behalf is recorded or discernible in the documents which deal with the negotiations and conclusion of these contracts. The payments made to Marlborough were not in the nature of bribery or the result of a secret arrangement between Sir Solomon and the General. They were made, as

[21] *Journal to Stella*, 23 January, 1712.
[22] Benjamin Disraeli, *Sybil*, Bk. IV, Ch. VII.
[23] 'Queen Anne's Army Contractors', *Jewish Chronicle*, 28 June 1889, p, 16.
[24] Peter Aldag, *Juden Erobern England*, Berlin, 1940, pp. 162–168.
[25] See pp. 38, 62   [26] See Chapter Six.

Sir Solomon testified, because, when taking over the contracts in 1707 from Antonio Alvarez Machado, Sir Solomon at the same time took over and adhered to the old prevailing practice, which was in force since the opening of the War of the Spanish Succession in 1702, five years before Sir Solomon became contractor. The Commissioners investigating the matter did not find Medina guilty of bribery or of any unlawful act. Thus, the suspicion of both Marlborough and Sir Solomon, created as a result of overlooking or disregarding these facts, is entirely unjustified.

Nor do I think that Sir Winston Churchill does justice to Medina when, in the standard biography of his ancestor, Marlborough, he says[27] that he either

> had some grievance about the payments made to him or he was gained to the Government interest. Whatever the case, he certainly framed his deposition in an injurious and misleading form.

I do not think that Medina had a grievance against Marlborough, for the latter, as far as was compatible with his office, supported Medina's applications and petitions for payments due to him by the Government. This support was called for when Medina's efforts to obtain payment failed. Marlborough in his communications could only confirm the truth that shortage of funds would endanger the campaign. And this is precisely what he indicated. A number of such communications by Marlborough are to be found in the files.[28] This sympathetic assistance by the General also comes to light in some of the letters written by him to Medina. Thus, he informed Medina on 2 August 1709:[29]

> What I have written already to my Lord Treasurer relating to your losses the last campaign will, I believe, be sufficient to induce his Lordship to give you all reasonable satisfaction.

Another instance is a letter dated 23 January 1711,[30] in which Marlborough wrote that 'I shall give my best assistance to your son'[31] in his endeavours to recover the arrears due. Two months later, on 19 March 1711, and only nine months before Medina's

[27] Churchill, *op. cit.*, IV, p. 483.
[28] For instance, *Treasury Out Letters* (General), XIX, p. 78; *Treasury Minute Book* XVIII, pp. 153–154.
[29] *Marlborough Dispatches*, IV, p. 707.   [30] *Ibid.*, V, p. 256.
[31] *i.e.* son-in-law.

testimony in the Marlborough investigation, the General wrote the following letter from The Hague to Harley, the head of the Government:[32]

> Sir Solomon Medina being come hither two days ago to represent the great difficulties he lies under for want of money to carry on the service, and having a letter by the last post from his son in London that 41,000L. was ordered him in tallies for this year's advance, he has taken the resolution of going over by this packet-boat to lay his case before the Lords Commissioners of the Treasury; and being very sensible of the great loss he sustained by his contract two years ago, I cannot forbear, in justice as well as in regard to the service, to recommend his pretensions to your favourable protection, that he may have at least part of his advance in ready money, and that his arrears may be considered at the same time, without which he protests to me it will be impossible for him to continue his deliveries to the end of the campaign.

Surely all this indicates that, far from having any grievance against Marlborough, Medina had every reason to be grateful for such valuable assistance.

There is also no evidence that in his testimony in the Marlborough matter, as Sir Winston Churchill writes, Sir Solomon was gained to the Government and thus influenced against the General. If Medina had any grievance in 1711—and it has been shown above[33] that he had very many—then this was directed against the Government for not paying the arrears due to him, thus causing him much harm and damage. He was badly let down by the Government, both before and after his testimony, and his grievance against it was so strong and he felt so bitter against it that four weeks after Marlborough's dismissal Medina 'declined contracting for the bread and bread waggons for the campaign of 1712'.[34] The authorities did not even try to persuade him to reconsider his decision, and appointed another contractor. This disposes of Sir Winston Churchill's suspicion. Marlborough himself did not deny the correctness of Medina's statements, and, on reading his deposition, wrote a letter to Lord Oxford on 10 November 1711 saying that 'my name was brought before the Commissioners of Accounts, possibly without any design to do me a prejudice.'[35]

I think this represents a fair estimate of Medina's attitude.

[32] *Marlborough Dispatches*, V, p. 274.
[33] See pp. 42, 48–56.   [34] *Treasury Out Letters* (General), XX, p. 131.
[35] Coxe, *op. cit.*, VI, p. 126.

While Medina's name is usually tied up with the dismissal of Marlborough, the much more worthy aspect of Sir Solomon's importance in connection with Marlborough's conduct of war is entirely omitted. Military supply is a dominant factor for an army and there is no doubt that its elasticity and good organisation contributed in no small measure to Marlborough's victories, as it contributed in other wars to the victorious commanders. It was under Marlborough that marked progress was made with the supply of bread for the armies, which was not only provided for the whole force but was for the first time in history delivered and carried in wagons.[36] History books pay high tribute to Lord Godolphin 'for the perfection of Marlborough's transport and supplies'.[37] As has been shown in a previous chapter,[38] Sir Solomon was responsible for this task. Therefore I think that when the military victories of Blenheim, Elixheim, Ramillies, Oudenarde, or Malplaquet are praised in tribute to British military bravery and genius, the 'Jew Medina' deserves an honourable place for his share in them.

MEDINA AND VOLTAIRE

The second event which brought Medina's name into the limelight of a wider public was Voltaire's London affair. In connection therewith, Wolf wrote that[39]

> the banking firm he [Medina] conducted in London, failed about twenty years after his death, and Voltaire, who was a creditor, avenged himself for his losses by his famous attack on the Jews.

This statement has to be corrected almost in its entirety. More than twenty years after Medina's death (he died in 1730), Voltaire lost some money through Abraham Hirschl in Berlin[40] (1750/1). He lost his money in London in 1726, when Sir Solomon was still alive; but, as will be seen, neither Medina nor a Medina firm was implicated.

[36] Brigadier Cecil Edward Ronald Ince, 'Military Supply' in *Chambers's Encyclopaedia*, XIII, p. 296.
[37] Herbert W. Paul, *Queen Anne*, London, Goupil, 1906, p. 85, See also Churchill, *op. cit.*, II, p. 76.
[38] See pp. 37–42.
[39] His papers in University College London.
[40] *Cf.* Wilhelm Mangold, *Voltaire's Rechtsstreit mit dem koeniglichen Schutzjuden Hirschel, 1751*, Berlin, 1905.

When Voltaire landed in England in June 1726, he brought with him a bill of exchange, and as he was not in immediate need of money he did not present it. On presenting it later for payment, he was informed that the man and the firm on whom it had been drawn had become bankrupt three days before; and the money was lost. It so happened that the King heard of his misfortune and sent him a sum, presumably a hundred guineas, to relieve him of his embarrassment.[41]

Graetz named Medina[42] as the Jewish capitalist who caused Voltaire's loss, and all subsequent historians—as far as they mention this event—took it over from him uncritically. This applies to Wolf, Dubnow, and several encyclopedias.

The assumption that Medina was the Jewish banker in question was based on the vivid description which Voltaire wrote in 1771, forty-five years after the event:[43]

> Messieurs, lorsque M. Medina, votre compatriote, me fit à Londres une banqueroute de vingt mille francs, il y a quarante quatre ans, il me dit que *ce n'était pas sa faute, qu'il était malheureux qu'il n'avait jamais été enfant de Bélial, qu'il avait toujours tâché de vivre en fils de Dieu, c'est-à-dire en honnête homme, en bon israélite*. Il m'attendrit, je l'embrassai, nous louàmes Dieu ensemble, et je perdis quatre-vingts pour cent.

This assumption was strengthened by Sieveking's discovery of Voltaire's letter to Thiriot,[44] his lifelong friend, of about the end of 1726 or the beginning of 1727. In this letter Voltaire said in his own English:

> ... Now my dear Tiriot after having fully answered to what you asked about English books, let me acquaint you with an account of my for ever cursed fortune. j came again into England in the latter end of july very much dissatisfied with my secret voiage into france both unsuccessful and expensive. j had about me only some bills of exchange upon a jew called Medina for the sum of about eight or nin thousand french livres, rekoning all. at my coming to London i found my damned jew was broken. j was without a penny, sick to death of a violent ague a stranger, alone, helpless. ...

In connection with this episode, Voltaire mentioned in 1776/7

[41] John Churton Collins, *Voltaire in England*, New York, 1886, pp. 202–203.
[42] Graetz, *History of the Jews*, XI2, 1900, p. 48.
[43] Voltaire, *Questions sur l'Encyclopédie*, VII (article 'Juifs').
[44] A. Forbes Sieveking, 'Voltaire in England', in *The Athenaeum*, London, No. 3380, 6 August 1892, pp. 194–195.

another Jew in London as the banker in question—da Costa:[45]

> Leur secrétaire (*celui des juifs*) me dit que je suis fâché contre eux à cause de la banqueroute que me fit le juif Acosta, il y a cinquante ans, à Londres: il suppose que je lui confiai mon argent pour gagner un peu de temporel avec Israël. Je vous proteste, Messieurs, que je ne suis point fâché: j'arrivai trop tard chez M. Acosta; j'avais une lettre de change de vingt mille francs sur lui; il me dit qu'il avait declaré sa faillite la veille, et il eut la générosite de me donner quelques guinées qu'il pouvait se dispenser de m'accorder. Comptez, Messieurs, que j'ai essuyé des banqueroutes plus considerables de bons chrétiens, sans crier.

Graetz, in an article,[46] but strangely enough not in his great *History*, points to the fact that Voltaire 'indulged in financial transactions with a Portuguese Jew, whom he sometimes calls *Medina*, sometimes *Acosta*'. And, speaking of Voltaire's antisemitism, Graetz reveals in the same article that 'every personality of the Old Testament reminded him [Voltaire] of Jews and thinking of Jews he thought of Acosta in London and of Abraham Hirschl in Berlin, who, as he believed, caused him the loss of money and reputation'.[47]

Why then has Graetz in his *History* attributed Voltaire's loss in London to Medina? No explanation can be found for this in his work. The problem has, however, been clarified by Gustave Lanson.[48] He first investigated Voltaire's letter to Thiriot, which was quoted above, because it was a contemporary document (dated about 1726 or 1727). Apart from Sieveking's publication in 1892, this letter was also published in 1905 by M. Hettier,[49] who, on examining the original, pointed to the fact that the words 'called Medina' were in black ink over an older faded text. This makes the reference to Medina very hypothetical. Apart from this, Lanson compared the letter to Thiriot with the texts of 1771 and 1776/7 referred to above. That comparison proved to him Voltaire's fading memory. While in 1726/7 Voltaire wrote about a loss of 8,000–9,000 francs, he stated in 1771 that he had lost 20,000 francs.

[45] *Un chrétien contre six juifs*, XI niaiserie (ed. Moland), XXIX, p. 558.
[46] Graetz, 'Voltaire und die Juden', in *Monatsschrift fuer die Geschichte und Wissenschaft des Judentums*, 1868, p. 162.
[47] Ibid. p. 173.
[48] 'Voltaire et son banqueroutier juif en 1726. Medina ou Dacosta?,' in *La Revue Latine*, 1908, VII, pp. 33–41.
[49] In *Mémoirs de l'Académie des Sciences, Arts et Belles-Lettres de Caen*, 1905.

But Lanson's main argument centres on the 'Jewish bankrupt' in London. This could not have been Sir Solomon de Medina, as he did not live in London at the time, but since 1702 had lived in Amsterdam and The Hague. As has been shown in this study, he was a wealthy man and was never bankrupt. Nor could it have been the firm of Joseph Medina & Sons, of Amsterdam and London, which was flourishing years after 1726[50] and never went into bankruptcy. Lanson established that in 1726 or 1727 there was only one bankruptcy of a Jew or a Jewish firm recorded in London—that of Anthony Mendes da Costa. This da Costa died on 24 June 1726,[51] about three weeks after Voltaire's arrival in London. A year later the *London Gazette* reported[52]

> In pursuance to an Order made by the Right Honourable the Lord High Chancellor of Great Britain, the Commissioners in the Commission of bankrupt awarded against Anthony Mendes da Costa, late of London, Merchant, intend to meet on the 19th of July next, at three in the afternoon, at Guildhall, London; when and where the creditors are to come to choose a new Assignee or Assignees of the Estate and Effects of the said Bankrupt, in the room of John Mendes da Costa, deceased.

There is no doubt that this is the banker and bankrupt in question. It is astonishing that a historian of Dubnow's calibre, twenty years after Lanson's evidence, still adhered to Graetz's version of 1875 regarding Medina.[53]

One can only guess why Voltaire, at the age of 77, referred to Medina in 1771. There was once a Medina who lent him money. This happened in 1738, when he borrowed money from a Dutch Jew named Medina.[54] But there is also the possibility that Voltaire, being a close friend of Sarah Jennings, Duchess of Marlborough, the great General's wife,[55] may have frequently heard Medina's name mentioned in a house where it had certainly not been forgotten that the latter's testimony largely assisted towards the Duke's dismissal. After so many years

[50] *Cf.* Medina's will and P.R.O. C.11.1570/30 of 1741.
[51] *Daily Journal*, No. 1702, 29 June 1726.
[52] No. 6592. From Tuesday, 20 June, to Saturday, 24 June 1727.
[53] Dubnow, *History of the Jews*, VII, p. 404.
[54] *Œuvres complètes*, ed. L. Moland, XXXIV, pp. 440, 449, 463, 464.
[55] Lytton Strachey, 'Voltaire and England', in *The Edinburgh Review*, 1914 (October), p. 399.

Voltaire may have exchanged the familiar name of the 'Jew banker' Medina for that of the 'Jew banker' da Costa. But this is pure hypothesis. The facts, however, are clear: the London incident contributed to Voltaire's anti-Jewish bias. But Sir Solomon de Medina had no part whatsoever in it.

APPENDIX A

The Identity of Solomon and Diego de Medina

As stated in the text, there is, curiously enough, no documentary evidence available about Medina's origins. Roth, basing himself on an old register of names, assumed that Medina originally came from Leghorn.[1] I have not been able to trace any corroborating evidence, though it must be pointed out that there were many Medina families well established in Leghorn whose family and business contacts with Amsterdam and London were very close indeed.

Mr. Wilfred S. Samuel drew my attention to his assumption that Solomon might have been identical with Diego de Medina. I found in Lucien Wolf's papers deposited at University College, London, a loose note which confirms Mr. Samuel's assumption and reads as follows: 'Diego de Medina was no doubt Salomon de Medina. See his petition for Denization, in my Denization papers.' In accordance with this note, I looked up Wolf's denization file, also deposited at University College, and found there a reference to Diego de Medina in connection with a document in the British Museum (Add. MS. 28074) only. This document, however, dated November 1672, represents the grant of denization, not the petition. The latter is not among Wolf's denization papers. I found the petition subsequently in the course of my research in the Public Record Office. It is for the first time referred to (but not quoted) in an entry dated 31 October 1672,[2] and extensively quoted in the grant of denization, which reads as follows:[3]

S.O. Docquet. Nov. 1672
A grant of denization to Diego Medina etc.

[1] Cecil Roth, *A History of the Jews in England*, Oxford, 1941, p. 283.
[2] *S.P.Dom.* Charles II, p. 105, 31 October 1672 (Whitehall).
[3] *S.P.Dom.* Entry Book 36, pp. 136/137.

Whereas Diego de Medina, Antonio Gomez Serra, Francisco de Lix and Jeronimo Fernandes de Miranda, Merchants, after their humble Petition informed Us: that in pursuance of Our gracious Declaration of the 12th of June last, he the first Diego de Medina came with his wife and family and goods from Amsterdam where he did live for several years unto Our City of London to settle himself under Our Government: And whereas their Petitioners Antonio Gomez Serra, Francisco de Lix and Jeronimo Fernandes de Miranda came from Amsterdam before Our gracious Declaration and settled themselves with their wives and familys in Our City of London and all of them humbly pray to have the benefit of Our Declaration and that is their purpose that Letters Patents may be granted unto them under Our Great Seale to make them free denizons of this Our Kingdom with a priviledge to import and export goods and Merchandizes paying no more nor other Customs for the same the inwards and outwards than Our naturall borne Subjects doe in like cases: We are graciously pleased to condescend to that their request, And it is accordingly Our will and pleasure . . . to make them free denizens. . . .

Whitehall. October ult. 1672.

Thus it was about four months after his and his family's arrival in England[4] that Medina was made a free denizen.

As stated, Wolf based his conclusion on another document, which is dated later than that quoted above, and reads as follows:[5]

November 1672.

Diego de Medina

A Grant from his Matie to make Diego de Medina, Antonio Gomes Serra Francisco de Liz & Jeronimo Fernandez de Miranda natives of France free Denizones of this his Maties Kingdom of England, to have receive and enjoy all Rights Privileges and Immunities, as other free Denizones doe, & to pay noe more nor other Customes for any Goods or Merchandizes by them imported or exported than his Maties naturall borne Subjects doe in like cases. Signifyed to be his Maties pleasure by Warrant under his Royall Signe Manuall.

Tho: Butler Dep^{ty}
C. Sickerstaffe

I have bin acquainted with this Docquett
Clifford

Dec. 9 1672
  Coms of Privy Seal
Another of ye same Contents for ye Great Seal.

This grant does not refer to the actual time of Diego de Medina's arrival in England, so that it was a hypothesis

---

[4] See pp. 1–2. [5] British Museum, Add. MS. 28074, p. 115 (p. 58).

APPENDIX A 77

rather than evidence on Wolf's part, only leading him to the identity of Diego and Solomon.

There is no documentary evidence of Solomon's arrival in England and his denization. In view of the wide ramifications of his business activities,[6] one cannot believe that he was not granted the privileges which were then vital for Jewish businessmen, and which he actually enjoyed. On the other hand, the documents refer to the arrival of Diego and his family in England and to his subsequent denization.

Diego's name is not mentioned at all in the Synagogue books. If he was a Jew—which is hardly to be doubted—this is very strange indeed, for he was a man of substance, who, as the documents quoted above reveal, brought his 'goods from Amsterdam' and 'imported or exported' merchandise. At that time, a Jew coming to London even temporarily on business was obliged to pay his *Imposta*[7] to the Congregation, but Diego is not recorded as having paid this, although he was not a visitor, for he was admitted to become a resident in London. Nor is he recorded in the Synagogue books in any other connection. On the other hand, Solomon is referred to as having paid his *Imposta* in 1670,[8] and from then onwards his name is continually mentioned in the Bevis Marks records, as has been referred to above in Chapter Three.

It is noteworthy in this connection that Jeronimo Fernandez de Miranda, mentioned in the petition for and in the grant of denization quoted above, is also referred to in the Synagogue books for the first time in 1670.[9] Both were—in accordance with the denization application—natives of France (Bordeaux)[10] and both were jointly honoured by the London Sephardim on Simchath Torah of 5431 (1670), when Medina was Hatan Bereshith and Jacob de Miranda Hatan Torah.[11] While Diego is recorded in that source as having arrived in England in 1672, Miranda is stated therein to have lived here already before the King's declaration, as is confirmed by the entry in *El Libro* above referred to.

Also, Francisco, the third denizen, known as Jacob Berahel, lived in England before the King's proclamation, as the

[6] See Chapters Four to Six.
[8] *Ibid.*, p. 47.
[10] Pat. Roll 24, Car. II, pt. 4.
[7] *El Libro* . . ., p. 5.
[9] *Ibid.*, p. 46.
[11] *El Libro* . . ., p. 48.

Christopher Dodsworth's Proceedings against the Exportation of SILVER by the JEWS and others.

Mr. *Cory's* Affidavit.

John Cory, of the Parish of S. *Clemens Danes,* in the County of *Middlesex,* Gent maketh Oath, That about the Month of *May* 1689. he this Deponent was desired by the Right Honourable the Earl of *Monmouth* to bring one Mr. *Michal Levy,* a Merchant and Jew to speak with him; which soon after this Deponent accordingly did, at his Lordship's House at *Parsons Green*: where, in this Deponent's hearing, his Lordship told the said Mr. *Levy,* That the occasion of his sending for him, was to acquaint him, That their Majesties wanted Mony, and that he believed the Jews to be a wealthy People, and could lend them a considerable Sum upon the Act of Parliament, at seven *per cent,* for carrying on the War against *France,* and that if ever they expected Favour from the present Government, then was the time to deserve it, by complying with their Majesties Occasions, or Words to that effect. To which Mr. *Levy* replyed, in this Deponent's hearing, That there was not above seventy or eighty Families in *England,* and of them not above seventeen or eighteen were Men of any considerable Estate; nevertheless he would use his Endeavour to serve their Majesties in what they desired; but nothing further was done in it. And this Deponent did soon after receive intimation, That the Jews had made Entry in the Months of *June* and *July* of one hundred forty eight thousand and two hundred Ounces of Silver, or thereabouts, for beyond the Seas, as by the Custome-house Books may appear, to which this Deponent refers himself. And this Deponent did soon after intimate the same in Writing to the Right Honourable the Lords Commissioners of the Treasury; who sent it inclosed to the Commissioners of the Customes: Who sent to this Deponent to speak with him upon the said matter; which he accordingly did, and averred the Entries of the said Silver. But the Commissioners were pleased to answer this Deponent, That Forreign Silver imported might be exported; and Mr. *Booth* in particular gave him an undue Reprimand for his endeavouring to serve their Majesties in such an important affair: So that it went beyond the Sea unsearched, as this Deponent verily believes.

Jurat vicesimo quarto Octobr.
Anno secund. Will. & Mar.   *John Cory.*
Rex & Regin. coram me
     N. Lechmer.

Mr. *Lawrence Swann's* Affidavit.

Lawrence Swann of the Parish of S. *Saviours Southwark,* in the County of *Surrey* Founder, maketh Oath, That this Deponent, at the Request of Captain *Leneve,* a Custome-house Officer, on *Saturday* the thirteenth day of this instant *September,* did go on board the *Soes-Dyke-Tatch,* to see some Bars or Pigs of Silver, amounting to about thirty thousand Ounces, and a quantity of Pieces of Eight, to the number of about five thousand; which by the said Captain were seized: And this Deponent saith, that upon view of the same, he informed the said Captain, That the quantity he then set aside was Bad, the whole whereof amounted to about sixteen thousand Ounces, as this Deponent was informed, and verily believes the same to be. And this Deponent further saith, That he, at the Request of the said *Leneve,* made an Essay of eleven Grains, which to the full value was not worth above four Shilling the Ounce, which should be worth five Shillings and two Pence the Ounce. And this Deponent verily believes the Pigs or Bars were cast in *England.*

Jurat decimo octavo die Septem.
Anno secundo Willielmi &   *Lawrence Swann.*
Maria Rex & Regina, coram me   Geo. Bradbury.

Mr. Attorney General's Opinion.

Whether divers sorts of Silver, imported and melted down together into Pigs, whereby the Quality of the whole Mass being altered, be not a Manufacturing of it? And whether it can then properly be called Bulloyn, or ought to pay Duty outwards?

It seems to me, that by reason of the Alteration it has undergone here, it doth not remain Forreign, Bulloyn, but may be liable to Duty outwards.

20 *Sept.*
1690.   *George Treby.*

Mr. Dodsworth's *Petition to the Lords Commissioners of the Treasury.*

To the Right Honourable, the Lords Commissioners of their Majesties Treasury; The humble Petition of Christopher Dodsworth *Merchant, sheweth,*

That of late Years the Exportation of Silver out of this Kingdom has been so great, that the Working Goldsmiths, the last Sessions of Parliament, petitioned for Redress thereof, and in their Petition and Paper they assert, That the Mill'd Mony of this Kingdom is usually melted down and exported, not only to their own Disadvantage, but to the Nation in general, (a Copy of which Petition and Paper are annexed).

That your Petitioner taking notice thereof, hath examined and found, That by the Entries at the Custome house *London,* from the third of *March* last to the 11th of this instant *September,* about 600000 Ounces of Silver hath been publickly ship'd off for *Holland* or other Forreign Markets.

That the Law gives leave for Exportation only of Forreign Bulloyn, and that he was informed if it were melted down here, 'twas an English Manufacture, and ought to pay Custome outwards; and being advised, that above 60000 Ounces was lately ship'd and shiping off, your Petitioner acquainted Captain *Leneve* and Mr. *Wright* two Officers of the Customes therewith, and went in person with them on board one of the Ships, where they found about 35000 Ounces, and that about 16000 Ounces was found mix'd Metal, not worth 4 s. *per* Ounce, when the Standard was 5 s. 2 d. and thereon concluding it an English Manufacture, the Officers seized it, as *per Affidavit.* annexed appears.

That soon after they acquainted the Commissioners of the Customes thereof, who drew up a *Quære* to be put to Mr. Attorney General (a Copy whereof is likewise annexed) and both Parties concluded to be guided thereby, as to point of Law; But Mr. Attorney's Opinion favouring the payment of the Customes, Sir *John Worden* declared against breaking an old Custome, and would not give any order for bringing that which was seized on Shore, and in fine discourag'd the whole Proceeding, so that the Officers are fearful to act any farther.

That

APPENDIX A

That the Cuſtome of the parcel now ſo lately ſhip'd amounts to 800 l. and for 6 Months paſt to 6000 l.

Your Petitioner therefore humbly prays your Lordſhips to take this into your Conſideration, both with regard to their Majeſties Revenue and the publick Good, and hear what may be ſaid by Counſel on both ſides, if to your great Wiſdom it ſhall ſeem meet; and that in the mean time ſuch Pigs of the ſaid Silver, as upon view ſhall be found to be caſt in *England* may be brought into the King's Warehouſe and there remain.

And your Petitioner ſhall ever pray, &c.

Mr Dodſworth's Petition to the Commiſſioners of the Cuſtomes.

To the Honourable Commiſſioners of their Majeſties Cuſtomes. The humble Petition of Chriſtopher Dodſworth Merchant, Sheweth,

THat your Petitioner having waited on the Right Honourable the Lords Commiſſioners of the Treaſury, about the Exportation of Silver, their Lordſhips have referred the conſideration thereof to your Honours, and in regard conſiderable Quantities of Silver have been ſhipt ſince the Seiſure made by Mr. *Leneve*, without examination whether manufactured in *England* or not, and that both Parcels, together with others, are ſtill on board ſome Ships in the River.

Your Petitioner humbly prays this Honourable Board either to cauſe all the Silver now on Board the ſaid Ships to be brought on Shoar to the King's Warehouſe, and there examined, or two able Workmen may be ſent on board with your Petitioner, to examine what Pigs of the ſaid Silver, are of Engliſh melting down and manufacturing; and ſuch as are ſo to ſecure in the ſaid Warehouſe, till the caſe be legally decided,

And your Petitioner ſhall ever pray, &c.

Mr. Dodſworth's Affidavit.

CHriſtopher *Dodſworth*, of the City of *London* Merchant, maketh Oath, That he this Deponent having on the 26th of *September* laſt paſt waited on the Right Honourable the Lords Commiſſioners of the Treaſury, with his Petition and Papers annexed about the Exportation of Silver, their Lordſhips were pleaſed to refer the Conſideration of the ſame to the Honourable Commiſſioners of their Majeſties Cuſtomes, to give their Opinion therein with expedition; which Reference being preſented to their Honours the 30th day of the ſaid Month of *September*, when Sir *John Worden* ſaid, they were of the ſame Opinion as before: Which this Deponent believes was, That no Examination ſhould be made whether the Silver in queſtion was of Engliſh Manufactory or not, nor that the Seizure made by Captain *Leneve* of part thereof ſhould have any effect. This Deponent further ſaith, That he waited again on the ſaid Commiſſioners of the Cuſtomes this preſent 2d of *October* 1690. and deſired to know if their Honours had given their Report, but Sir *John Worden* anſwered, They were extream buſie and could not yet give their Opinion: And this Deponent verily believes, the Veſſels on which the ſaid Silver is ſhip'd, are now ready to depart.

Jurat 2. *Octobris* 1690.
coram me Geo. Bradbury.   *Chr. Dodſworth.*

They are gone with about 110000 Ounces of Silver, the Cuſtome whereof amounts to 1375 l.

Copy of the Lords Commiſſioners of the Treaſury's Letter to the Commiſſioners of the Cuſtomes.

Gentlemen,

WHereas the Lords Commiſſioners of their Majeſties Treaſury referred to you, the 27th of *September* laſt, the Petition of *Chriſtopher Dodſworth*, directing you to make your Report thereupon with all convenient Speed; but you having not yet ſent us the ſame, their Lordſhips direct you forthwith to your Report upon the ſaid Petition, which or the ſence of Mr. *Jepſon* is ſignified to you by, Gentlemen,

Treaſury Chamber   *Your moſt humble Servant*
Octob. 7. 1690
                                           William Lowndes.

The Ships went away on *Saturday* the 4th of *October*.

Mr Dodſworth's Petition to the H. of Commons

To the Honourable the Knights Citizens and Burgeſſes in Parliament aſſembled, the humble Petition of Chriſtoph. Dodſworth *Merchant, Sheweth,*

THat your Petitioner having for ſome Months paſt taken notice of the unuſual ſhipping for *Holland* great quantities of Silver, by the Jews and ſome others, and alſo being acquainted that the Working Goldſmiths had, the late Seſſion complained thereof in this Honourable Houſe, aſſerting that the Mill'd and other heavy Coyne of this Kingdom was melted down, occaſioned him to enquire, Whether the Silver thus ſhip'd out was really Forreign Silver, as entred in the Cuſtome-houſe, and ſo by Act of Trade 15 *Car. 2*. to paſs out free, or Silver melted down here, and ſo an Engliſh Manufacture, which by the Act of Tonnage and Poundage 12 *Car 2.* continued *anno primo Gul. & Mar.* ought to pay Cuſtome *ad valorem.*

That on *Saturday* the 13th of *September* laſt your Petitioner, together with Captain *Leneve* and Sir *Wm Joht*, Officers of the Cuſtomes, and Mr. *Sweete* an Artiſt, went on Board one of the Ships outward bound, where they found about 25000 Ounces of Silver, which Engliſh caſting, about 16 00 Ounces was found on the Teſt to be worth not above 4s. *per* Ounce, and conſequently a mix'd Metal; this th. Officers ſeized, and on *Monday* following acquainted the Commiſſioners of the Cuſtomes therewith: Who did not only diſcourage that Seizure, but ſuffered about 700 0 Ounces more to be ſhip'd off, without Examination whether Engliſh or Forreign, and without Payment of Duty, though Mr. Attorney General was of Opinion, That ſuch as was altered in *England* was lyable to the Duty.

That thereon your Petitioner laid the caſe before the Lords Commiſſioners of the Treaſury, who refer'd him to the Commiſſioners of Cuſtomes, from whom no Report came till the Ships were gone. Copies of all his Papers relating thereunto are ready to be produc'd.

That for 16 Months paſt about 1400 00 Ounces of Silver hath been thus exported (which ſome Cuſtome Officers ſay is more than was in 16 Year before, except to the *Eaſt-Indies*) and the Cuſtome thereof unpaid amounts to 17500 l. ſterling.

Your Petitioner therefore humbly prays this Honourable Houſe to take the Premiſes into Conſideration, both with regard to the King's Revenue and the Publick Good, and order thereon as to your Honours great Wiſdom it ſhall ſeem meet,

And your Petitioner ſhall ever pray, &c.

The Names of the Exporters of Silvers, as by the Cuſtomehouſe Entries appear.

Jews. ALvaro de Coſta, Jacques Gonſalez, Alphonſo Rodrigues, Antonio Rodriguez Marques, Joſeph Bueno, Antonio Gomezſera, John Fermaeo, Caleb Papall, Solomon Levi, Peter Henriquez, Peir Henriquez, Symon Garo, Elias de Mevira, Paſ. Levi, John De Lean, Antonio Correa, Joſeph Meza, Jaſper Francico, Joſeph Paper, Simon Byrloam, Joſias Malez, Symon Francia, Moſes Carroen, George De Lapo, John dopont, Joſeph Levi, Jaſper Perrero, Joſeph Marquez, James Carroon, Mordecay Iſaac, Deigo de Medina, Oder Pomra, Antonio Rodrigues, Peter Perrera, Iſaac Gomez, Peter de Faxaia, William de Coſquet, James Zibezherio.

Dutchmen. John Scopens, John Vanhine, John Vinderſpoo, G, Vanvulgh, Derrick Symons, Solomon Blockar, John Vanderhome, Geo. Vanderhoon

Engliſhmen. William Roberſon, Peter Jackſon, John Johnſon, Geo. Smith, John Palmer, John Bryan, Th. White, Alexander Pryer, John Valleurme, John Thompſon, Peter Ramsay, Tho. Allen, John Sweecapie, Peter Newaeash, Peter Hartom, Antho. Stone, Peter Bull, John Sheeriff, Joſeph Harley, John Phillips, Alexander Goodman, Joſeph Bant, William Harden, Will. Sneding, Walter Nahs, James Six, Henry, Stephen Evans.

denization grant confirms. As a matter of fact he was one of the signatories of the *Ascamot* of the London congregation as far back as 1663.[12]

There is still the gap of two years left unbridged between the first reference to Solomon in 1670 and Diego's arrival in England in 1672. However, it can be assumed that Solomon arrived in England from Amsterdam in 1670 and paid his *Imposta* for the last six months of that year (5430), as everyone was obliged to do whether resident or visitor. Payment 'for the last six months' was in accordance with the rules made 'eight days before Rosasana',[13] which in 1670 fell on 15 September. Having thus arrived in London, he stayed over the Festivals, and was, as stated above, elected Hatan Bereshith and called up on 7 October.[14] Apparently this was not a short visit, because Solomon is recorded as having also paid his *Imposta* in September 1671 for the whole year 5431.[15] Significantly, Solomon did not pay any *Imposta* for 1672. He may have returned to Amsterdam, and when on 12 June 1672 the King's proclamation was issued[16]

> for encouraging the subjects of the United Provinces of the Netherlands to transport themselves, with their estates and to settle in this His Majesty's Kingdom of England,

he may have availed himself of this chance and brought his family to England to settle there permanently. It was quite customary among Jews of Spanish or Portuguese origin then to travel with their official papers or passes made out in their legitimate non-Jewish names, while among their people and in synagogal life they used their Hebrew names. It is thus possible for Solomon de Medina to have travelled with a pass made out in Diego's name, and accordingly to apply for denization in this official name.

Further support for the identity of Diego and Solomon: His first grandson was named Solomon alias Diego,[17] obviously to perpetuate both names of the illustrious grandfather.

There is, finally, one further documentary proof available, for the photocopy of which (see pages plates 78, 79,) I am grateful to Dr. Cecil Roth. This establishes conclusively the identity

[12] *Ibid.* p. 14.
[13] *Ibid.*, p. 5.
[14] *Ibid.*, p. 48.
[15] *Ibid.*, p. 56.
[16] *S.P.Dom.* Vol. 71, p. 210.
[17] Sir Solomon's will, Appendix C.

APPENDIX A 81

of Diego and Solomon de Medina. In September 1690, Christopher Dodsworth's petition against the export of silver from England by the Jews was investigated by the Attorney General and by the Lords Commissioners of the Treasury. An enclosure among the documents contains a list of Jewish exporters of silver 'as by the Customhouse Entries appear.' In the list 'Diego de Medina' is mentioned. It is quite obvious that this can only be Solomon, who, in 1690, was exporting silver and is the only Medina then to have done so.[18]

[18] [This statement by the author appears to ignore the fact that in Christopher Dodsworth's list, as shown in the reproduction of his 'proceedings' on another page, an 'Elias de Medina' is also mentioned among the Jews he charges with illegally exporting silver. But although Dr. Rabinowicz does not mention Elias de Medina in this book, he was not unaware of the occurrence of the name. An Eli (presumably the same as Elias) de Medina is mentioned in *Misc. JHSE*, Part I (1925), page xxviii, in a list compiled by I. Abrahams of Jews who had been granted passes for travel abroad between 1689 and 1696, as they were considered to represent no danger of disloyalty to the Government of William III in respect to Jacobite intercourse with France. Eli de Medina was given his pass on 14 March 1692. In response to my query, Mrs. Judith K. Tapiero, daughter of Dr. Rabinowicz, searched again through his papers and has provided the following note, which goes some way towards explaining why he omitted Eli (Elias):

'The puzzle in Appendix A concerning Elias/Eli de Medina is indeed baffling, but I do not think it changes my father's well-documented belief that Solomon and Diego were one and the same. There are too many other proofs substantiating his claim. My father would never have left such a loose end, one that would leave him open to almost certain criticism. Eli also seems to have been a problem to Lucien Wolf, as he says in his papers; there are a few places where Eli is mentioned but there is nothing to indicate his identity. It is apparent that Eli was the Jewish name of one of the Medinas. Wolf's papers reveal: "I have several instances of Francisco who took Elias for their Jewish name. There is a good precedent for identifying Eli Medina as Francisco Medina [the father of Solomon]. My reason for believing that he was the father of Solomon de Medina is that the latter's elder brother Joseph called his son Francisco." My father states, on the other hand, "Eli is the only de Medina in the Amsterdam list of married members for 1675 and he is also recorded as having received a pass in London in 1692 (*S.P.Dom.* Vol. 82, p. 179; H.O. Letter Book 2, p. 175). However, Wolf's statement only indicates the fatherhood of Francisco . . . but not the identity with Eli, which would have to be proved."'

—Ed.]

APPENDIX B

Joseph de Medina's Family

The life story of Sir Solomon de Medina's brother, as far as it concerns the year and place of birth as well as his youth, is as impenetrable as that of Sir Solomon himself. Joseph Aaron de Medina was older than Solomon. His eldest son, Moses, was born in 1674 or 1675,[1] but Solomon's daughter was already born when she came to England with her parents in 1672. Joseph was either born shortly before his parents left Bordeaux and settled in Holland, or soon after their arrival there before or during 1649. It was customary among Sephardi Jews to marry at an early age, and if this custom can serve as even remote evidence, Joseph might have been born in Holland in about 1652 and was approximately 22 years of age when his son Moses was born. Joseph's wife was Rachel,[2] whom he brought along when he visited London in 1692.[3] It is only by way of a passing reference in a court deposition that we learn of Joseph's death before 1727.[4]

In the text we have already met Joseph as head of the Amsterdam firm Joseph Medina & Sons, as well as collaborator of Sir Solomon, and dealt with this fully in the narrative.[5]

Joseph de Medina was neither as successful as his illustrious brother nor does he appear to have been able to make much headway in his early business efforts, as can be gathered from the will of Diego Rodrigues Marques, of London, dated 10 November 1675:[6]

> ITEM Besides all that is mentioned in this my Will I doe declare that I have an account with Joseph de Medina of Amsterdam and for Ballance thereof he oweth me about £40 Sterling. Item I doe further declare that I doe owe to Salomon de Medina about £30 a little more or lesse which I did not pay him by reason of a difference which I have with his brother Joseph de Medina.

[1] See pp. 33, 84.
[2] *S.P.Dom.*, Vol. 82 (1691–1692), p. 389.
[3] *Ibid.*
[4] P.R.O. C.11. 1570/30 (1741).
[5] See pp. 26, 33–35.
[6] *P.C.C.* 113 Reeve.

APPENDIX B

The poor financial situation of Joseph de Medina's family is also discernible in a statement made in 1712 by Sir Solomon, in which he said[7] that Moses de Medina, Joseph's son, 'came bare to my house and because he was a Branch of my family gave him my Daughter in marriage with a suitable portion'.

Joseph Aaron and Rachel de Medina had four children, one daughter and three sons.

(1) Isaac de Medina, 'Senior',[8] was born in 1680[9] and in 1702, after the formation of Joseph de Medina & Sons, became a partner in his father's firm.[10] In addition to this, he later speculated independently in securities.[11] He is also listed as a shareholder in the Utrechtsche Compagnie.[12] He married Ester Nunes Miranda, from London,[13] on 7 September 1703. The last entry I traced about him was in a court deposition in 1741,[14] when he was 61 and active in business. The suffix 'Senior' distinguished him from his nephew Isaac, the son of his brother Moses.[15]

Isaac Senior had a son Aaron (named after his grandfather Joseph Aaron de Medina), but he died at an early age between 3 June 1727 and 1 September 1728.[16]

He also had a daughter Rachel (named after her grandmother). As will be seen below, Moses, Isaac Senior's brother, also named one of his daughters Rachel. Isaac's daughter married her cousin Solomon alias Diego, the son of Moses de Medina, on 28 Iyar 5483 (2 June 1723) in London.[17] She died in London and was buried at Mile End on 19 Ab 5514 (7 August 1754), in row 16, No. 45.[18]

[7] *S.P.Tr.* T.1.154/10A.
[8] Medina's will (Appendix C); P.R.O. C.11.1570/30.
[9] Municipal Archives of Amsterdam, Arch. No. 703/170.
[10] P.R.O. C.11. 1570/30; *cf.* also p. 34.
[11] Bloom, *op. cit.*, p. 189, n. 59.
[12] Jacob Zwarts, *Hoofdstukken uit de Geschiedenis der Joden in Nederland*. Zutphen, W. J. Thieme, 1929, appendix pp. 7–11.
[13] Municipal Archives of Amsterdam, Arch. No. 703/170.
[14] P.R.O. C.11. 1570/30.   [15] Isaac de Medina's (jun.) will, Browne 17.
[16] Medina's will (Appendix C) of 3 June 1727 mentions Aaron 'the Innocent and incapable Son of his Testators Nephew', but in the codicil of 1 September 1728 he is reported to have died since the date of drawing the will.
[17] *Bevis Marks Records* II, p. 74, No. 303.   [18] Burial Register.

(2) Deborah, who was named after her grandmother, Gracia Pereira, married Abraham de Solomon Mendes, on 2 Tebet 5457 (4 December 1697),[19] in London.

(3) Solomon Hiskia alias Francesco is dealt with further below in this appendix.

(4) Moses de Medina was 'alien born'[20] in Amsterdam[21] in 1674 or 1675.[22] He was made a free denizen on 29 November 1697.[23] He married Selomoh de Medina's daughter Deborah, and the wedding took place on 5 Heshvan 5453 (15 July 1692). Her dowry was £2,000 and £1,000.[24] It is probable that Moses' parents came over to England from Holland for the wedding, for Joseph and Rachel de Medina and their son Isaac left for Holland on 29 July 1692 via Harwich.[25] Moses' own business activities, and those in collaboration with his father's firm, as well as with his father-in-law, are dealt with elsewhere.[26] He gradually emerged among the London Jewish community as an important member in his own right. But this was not before Selomoh's departure from England in 1702. In the following year Moses signed, together with Antonio (Jacob) Gomez Serra and Andrew (Abraham) Lopez, a petition on behalf of the Jewish inhabitants of Jamaica, showing[27] that 'for the better peopling of JAMAICA an Act passed in 1683 empowering the Governor to grant letters of naturalization to all foreigners who should settle there. Under this act the Jews have settled there and taken out such letters. During the Governorship of SIR THOMAS LYNCH, LORD VAUGHAN, LORD CARLISLE, the DUKE OF ALBEMARLE, and LORD INCHIQUIN invidious persons tried to have them more heavily taxed than other inhabitants but could not induce any of the Governors to break through the law and infringe the

---

[19] *Bevis Marks Records* II, p. 64, No. 45.
[20] His denization grant in *S.P.Dom.* Entry Book 347, p. 104.
[21] P.R.O. C.24/1397 (Town Depositions).
[22] P.R.O. C.24/1338; C.24/1397; C.24/1418; C.24/1435; C.24/1443; C.24/1462; C.24/1467 (Town Depositions).
[23] *S.P.Dom.* Entry Book 347, p. 104; and *S.P.Dom.* Vol. 86B, 1697, p. 491.
[24] Moses Gaster, 'The Ketuboth of Bevis Marks', *Misc. JHSE*, II, p. 76f., No. 13.
[25] *S.P.Dom.* Vol. 82 (1691-1692), p. 389; H.O. Letter Book 2, p. 314.
[26] See pp. 24, 26, 33-35, 42, 52.   [27] *S.P.Dom.* 26 February 1703.

petitioners patent, but in Sir William Buston's Governments they were taxed separately and more highly than the others. They appealed to the late King who directed that this should not be done. Nevertheless by a recent act of the Assembly a special tax of £2,500 is imposed on the Jewish nation (over and above the sum of £17,808 to which they contribute as the other inhabitants) with a penalty of £500 in case of non-payment. Ruin, if not relieved. There are only 50 families and of these only 12 are in a position to contribute the extraordinary tax. Pray for relief.' This petition was successful.[28]

Moses de Medina became prominent in the famous controversy about whether or not Haham David Nieto, in his sermon on the 'Divine Providence', had exhibited heretical tendencies inspired by the pantheistic teachings of Spinoza.[29] It was Moses who informed the Gentlemen of the Mahamad in March 1704 that one Joshua Zarfatti had told him that he declined to enter the house of David de Avila, since he refused to be under the same roof as the Haham, considering him to be a heretic.[30] When, after long delays, Haham Zevi of Altona agreed to decide (together with two coadjutors) whether Nieto's sermon was conformable to the Law or not, it was Moses de Medina who, on 29 Tamuz 5465 (Tuesday 21 July 1705), forwarded the documents to Altona.[31] Moses was then Gabay, and for the first time held high office in the Synagogue.[32] The decision, which was in favour of Nieto, was published in Hebrew and Spanish in 1705 and contained the superscription: 'To the Illustrious from his good name and reputation [sic!], Moses de Medina, Gabay of the Holy Congregation of Shaar a Shamaim of the great city of London, which God preserve. Amen.'[33] The decision was republished in 1712 and 1716. On all these occasions, the superscription for Moses de Medina is also reprinted.[34]

[28] *S.P. Colonial America and West Indies*, Vol. XC, p. 446. See also Frank Cundall, N. Darnell Davis, and Albert Friedenberg, 'Documents Relating to the History of the Jews in Jamaica and Barbados in the time of William III', *PAJHS*, No. 23, 1915, pp. 25–29.

[29] This controversy is exhaustively elaborated by Israel Solomons, 'David Nieto and some of his Contemporaries', *Trans.JHSE*, XII, pp. 1–101.

[30] *Ibid.*, p. 10. [31] *Ibid.*, p. 16. [32] Bevis Marks Archives.
[33] Solomons, *op. cit.*, p. 66. [34] *Ibid.*, pp. 67, 68.

Twenty years later, in 1724, Moses de Medina was President of the Synagogue when another case caused uproar within the community. Dr. Jacob de Castro Sarmento was accused of causing the imprisonment of many Marranos in Beja (Portugal), but the Elders of the Synagogue, examining the case under Moses de Medina's Presidency, found him not guilty and publicly announced their decision on 24 Iyar 5484 (6 May 1724).[35]

As the Bevis Marks Archives show, Moses de Medina served as Parnas in 5469 (1708/1709), 5474 (1713/1714),[36] and 5478 (1717/1718). He was elected again in 5491 (1730), but declined, like his father-in-law almost thirty years previously,[37] and like Sir Selomoh he was fined. Moses' health was very poor by then,[38] and six months later he died.[39]

Moses de Medina owned a 'Copiehold house, garden and premises' at Richmond,[40] and a 'leasehold Estate' in Bevis Marks,[41] which perhaps included the house in Bury Street where he lived with his family.[42] The house in Bury Street was near Leadenhall Street.[43] From 1714 to 1722 he lived in Billiter Square,[44] Fenchurch Street.[45]

Until his death he was an important factor within the Jewish community and in the business world of England and Holland. He was a member of the London Royal Exchange[46] and from 1707 to 1730 a dealer in the Exchange Alley.[47] On some occasions business disputes were brought before him,[48]

[35] *Ibid.*, p. 84; see also Cecil Roth, *A History of the Marranos*, Philadelphia, Jewish Publication Society of America, 1941, p. 386.
[36] Also recorded in P.R.O. C.24/1338 (Town Depositions).
[37] See p. 9.   [38] Ester d'Azevedo's will, Isham 150 (last part).
[39] *Ibid.*   [40] His will, Isham 61 (1731).
[41] *Ibid.*   [42] *Ibid.*
[43] P.R.O. C.24/1462 (Town Depositions).
[44] P.R.O. C.24/1338 (Town Depositions).
[45] P.R.O. C.24/1345 (Town Depositions).
[46] P.R.O. C.24/1462 (Town Depositions).
[47] P.R.O. C.24/1443 (Town Depositions).
[48] A disinterested arbitration in 1720 landed him in some litigation. See P.R.O. C.11/1419/31 (Document No. 2, 'The Severall Answer of Moses de Medina', 8 February 1720); C.11/1768/24 (Document No. 2, 'Bill of Complaint', 14 May 1720); C.11/1992/30 (Document No. 1, 2 July 1720); C.11/1996/34 ('Bill of Complaint of Jacob Lopez Henriques of London Jeweller', 14 December 1721; and 'The Answer of Moses de Medina'

APPENDIX B

which proves the esteem in which he was held among his compatriots.

He died on 20 March 1731,[49] and a contemporary magazine recorded his death as that of a 'noted Jew'.[50] He was buried at Mile End, row 22, on 15 II Adar 5491 (23 March 1731).[51] His wife, Deborah, Sir Selomoh's daughter, died in May 1746, and was buried next to her husband on 5 Sivan, 5506 (24 May 1746).[52]

While Selomoh de Medina had only one child, his daughter Deborah, the wife of Moses, gave birth to nine children. They were four sons: Solomon, Aaron, Isaac, and Abraham, and five daughters: Ester, Rachel, Ribca, Abigael, and Jochebeth.

(1) Solomon was born on 14 September 1694[53] and circumcised a week later, on 21 September.[54] They named him Solomon after his grandfather Selomoh de Medina. Forthwith he was known as Solomon alias Diego. I referred to this in Appendix A as one of the proofs of the identity of Selomoh and Diego de Medina. In later years he was known as Solomon de Medina Mosesson[55] (this stands for: de Moses), and conducted his business in London under that name.[56] He lived in Bevis Marks.[57] He traded in tobacco and gold,[58] and apparently in other commodities also. One can assume that he may have gained his business experience from his own father,[59] and one

2 April 1722). See also P.R.O. C.11/2026/13; another arbitration in P.R.O. C.24/1462 (Town Depositions); C.24/1467 (Town Depositions); another in P.R.O. C.24/1397 (Town Depositions).

[49] Ester d'Azevedo's will, Isham 150 (1731).
[50] *Gentleman's Magazine*, March 1731, p. 126.
[51] Burial Register, Beth Holim.   [52] *Ibid.*
[53] Baptismal register of the Church of St. James, Westminster (*i.e.* St. James's Church in Piccadilly). Mr. Wilfred Samuel informed me that this registration signifies that Moses de Medina paid a fee to the Clerk or the incumbent so that the parents could have an official record of the birth of their son in England. It does not mean he was baptised.
[54] *Ibid.*   [55] Selomoh's will (Appendix C); P.R.O. C.11/807/2.
[56] *The Decree In the Case of Solomon de Medina Mosesson and Company, Merchants in London and Roderigo Pacheco, Jacob De Lara and Manuel da Costa, New York, 1728.* Dr. Cecil Roth suggested that as 'de Moses' he was called Mosesson in some Dutch records, and that in error this name found its way into the English records; it was not his 'surname'.
[57] P.R.O. C.24/1433 (Town Depositions).
[58] *The Decree in the Case* . . .   [59] P.R.O. C.11/1992/30.

cannot doubt that a close business relationship existed until the latter's death.[60] He also was a shareholder in the Utrechtsche Compagnie.[61] At one time his brother Abraham was also a member of his firm.[62]

Solomon Mosesson married his cousin Rachel, the daughter of Isaac de Medina (Senior) on 28 Iyar 5483 (2 June 1723).[63]

Solomon Mosesson seems to have been a gambler, which necessitated his grandfather's intervention. We find in Selomoh's will the following illuminating paragraph: 'ffurther the Appearer declared that Whereas at two several times he had assisted his said Grand Son Salomon de Medina Moses Son once with ffour hundred pounds Sterling and once with Three hundred pounds Capital South Sea Annuityes and that he is not in Condition to make Restitution to the Appearer for the same he the Appearer doth therefore hereby remit and discharge the same unto him and it is accordingly his will that the same shall not be required or demanded of him but nevertheless with this express condition that his said Grand Son from now and henceforth shall be obliged sincerely and exactly to perform the solemn promise made to the Appearer of never to play or enter into directly or indirectly any Game or Wagering wherein he within the Space of one Month should win or loose more than the value of an English Crown or three Guilders Hollands money in such wise that if he comes to act contrary to this promise and that he shall be convicted by his ffather or by his Brothers in an honest and lawful Wise within the Space of one Month of having at one Game or Wager directly or indirectly won or lost more than the value of an English Crown or three Guilders Hollands money it is Appearers Will that in such case there shall be imputed to him as well the four hundred pounds Sterling as the three hundred pounds Annuityes and that the value of this Seven hundred pounds shall be discounted and deducted from the Seven hundred and fifty pounds Bank Stock here before stipulated for his use so that after the decease of both the Appearers the said Sum of ffour

---

[60] *Acts of the Privy Council—Colonial Series III (1720–1745)*, p. 459, the appeal to the case referred to in footnote 56.
[61] Zwarts, *op. cit., loc. cit.*
[62] *Acts of the Privy Council, loc. cit.*
[63] *Bevis Marks Records* II, p. 74, No. 303. About Isaac see p. 83.

and three hundred pounds shall come to the Benefit of his two Brothers Isaac and Abraham de Medina.'

Solomon Mosesson died before July 1741.[64] His wife, Rachel, was still alive in 1743.[65]

(2) Aaron was another son of Moses and Deborah de Medina, named after the former's father: Joseph alias Aaron de Medina. He died at an early age, and he was buried at the Mile End Cemetery, row 9, on 28 Adar 5469 (10 March 1709).[66]

(3) Isaac, another of Moses de Medina's sons, has been mentioned above in connection with his uncle, Isaac de Medina Senior. He lived for some time 'beyond the Sea',[67] and died in 1739.[68] He does not mention a wife or children in his will, and it may be assumed that he was not married.

(4) Abraham, the fourth son of Moses and Deborah de Medina, married Sarah de Joseph Israel Sequeira[69] on 24 Kislev 5492 (23 December 1731), a few months after his father's death. He seems to have been poor, and left a negligible amount to his wife and children.[70] He lived in the parish of St. John Hackney,[71] died there, and was buried at Mile End, row 28, No. 5, on 18 Tishri 5535 (23 September 1774).[72] His wife, Sarah, 'died at her son's house at Stoke Newington, far advanced in years', on 10 November 1784.[73] She was buried on the following day next to her husband in row 28, No. 6, at Mile End.[74]

Abraham and Sarah de Medina had five children, one daughter and four sons.

(a) Sarah, born in 1734,[75] died unmarried[76] on 9 July 1787.[77]

(b) Moses (named after Abraham's father), who died in 1740 at the age of seven.[78]

(c) Joseph (named after Abraham's paternal grandfather), born in 1735.[79]

(d) A son who died in infancy (apparently before he was

[64] P.R.O. C.11/807/2.   [65] P.R.O. C.11/1584/18 (second document).
[66] Burial Register, Beth Holim.   [67] Moses' will, Isham 61.
[68] His will, Browne 17 (Probate).
[69] *Bevis Marks Records* II, p. 82, No. 501.   [70] His will, Bargrave 373.
[71] *Ibid.*   [72] Mile End Burial Register, p. 878.
[73] *Gentleman's Magazine*, November 1784.
[74] Mile End Burial Register.   [75] P.R.O. C.11/807/2.
[76] Tombstone inscription at Mile End, row 33, No. 6.   [77] *Ibid.*
[78] P.R.O. C.11/807/2.   [79] *Ibid.*

named) was buried on 24 Tebet 5498 (16 January 1738), row 2, No. 33, at Mile End.[80]

(e) Solomon (after Abraham's maternal grandfather, Selomoh de Medina), who, less than two months after his father's death, on 13 November 1774, married a non-Jewess, Elizabeth Coulrick, of St. Peter le Poor S. Lie.[81] Abraham, his father, knew the bride, who witnessed his will.[82] Sarah de Medina, Solomon's mother, lived with her son until her death on 10 November 1784.[83] Solomon, born in 1744,[84] was a stockbroker, whose offices were at Threadneedle Street[85] and later at 40 Basinghall Street, and was admitted as broker in 1777.[86] He also lent money on the basis of an annuity on the borrower's life.[87] He died in April 1823 at the age of 78, and was buried at Southampton[88] on 18 April. His wife, born in 1754, died at Stoke Newington on 23 September 1804[89] at the age of 50, and was buried on 29 September.[90]

Solomon and Elizabeth de Medina had eight children. Apart from the first two enumerated below (a and b), all the others are recorded in the Register of Baptism in the Church of St. Mary at Stoke Newington.[91]

(a) Ferdinand de Medina was born in 1775, and died on 6 November 1825 (aged 50). On 2 February 1803, he married Mary Mawton at Queen Street Chapel, Bath. Mary was born in 1777 and died on 29 December 1845 (aged 68).[92] They had three children: Eliza, who married Edward Headland, M.D., on 26 September 1826. She died on 2 December 1888 (aged 82);

---

[80] Mile End Burial Register, where his death is recorded 'Filho de Ab^m de Medina'.

[81] Marriage Register, St. Benet Fink, London. Recorded in Sir Thomas Colyer-Fergusson, Collection of Jewish Pedigrees (MS. bequeathed by Sir Thomas to the JHSE).

[82] Bargrave 373.   [83] *Gentleman's Magazine*, November 1784, p. 878.

[84] Colyer-Fergusson Collection.   [85] P.R.O. C.12/570/16.

[86] Dudley Abrahams, 'Jew Brokers of the City of London', *Misc. JHSE.* III, p. 91.

[87] P.R.O. C.12/570/16.

[88] All Saints Southampton. Burial Register 1823, No. 522 (Catacomb 63), in Colyer-Fergusson Collection.

[89] *Gentleman's Magazine*, September 1804, p. 891.

[90] Stoke Newington Register of Burials, in Colyer-Fergusson Collection.

[91] All the following references in the Colyer-Fergusson Collection.

[92] *Gentleman's Magazine*, December 1845, p. 259.

her husband died on 8 December 1869 (age 66). Their descendants include, *inter alia*, Harold Spender, J. A. Spender, Stephen Spender, the great poet, Sir Henry and Lady Fletcher, etc. Ferdinand de Medina's second daughter was Marianne, who married the Rev. John Swete, D.D., on 2 January 1838. He died on 7 September 1869 (aged 82). Among their descendants was Professor Henry Barclay Swete, the Biblical and patristic scholar of Cambridge. Ferdinand's only son was Ferdinand de Medina, born on 5 April 1805, who died on 28 March 1836 (aged 30).

(b) Charles Smith de Medina, of Hill in Hillbrook, Southampton, died on 10 August 1828.

(c) Louisa Charlotte, born in 1782, and baptised on 7 August 1782, married John Brenchley, of Lincoln's Inn Fields, on 1 January 1812 at St. George's, Bloomsbury.

(d) Solomon de Medina, born on 29 September 1783, and baptised on 30 October 1783, married Sophie Normansell on 10 May 1813 at Marylebone Church. Solomon died on 21 April 1820 (aged 36).[93]

(e) Elizabeth, born on 30 January 1788, and baptised on 15 March 1788, died on 7 January 1802 (aged 13),[94] and was buried at Stoke Newington on 13 January 1802.

(f) Sarah, born on 1 August 1789, was baptised on 16 September 1789. I cannot trace her husband's name.

(g) Caroline Sophie, born on 10 May 1793, and baptised on 13 June 1793, was married twice: (1) On 2 April 1802, at St. George's, Bloomsbury, to Edward Christopher Bryant, of Southampton. (2) On 30 March 1837, at Kew, to the Rev. Thomas Tunstall Haverfield. Caroline Sophie died on 8 August 1875 (aged 82), her second husband on 3 May 1866 (aged 76), and they were buried at Kew Cemetery. Among their descendants was F. Haverfield, of Oxford, the great authority on Roman Britain.

(h) Amelia, born on 14 August 1797, and baptised on 2 February 1789, married Henry Walker, of St. Laurence, Southampton, at All Saints in Southampton on 4 April 1822.

This shows that the Medina family merged through marriage

[93] *Gentleman's Magazine*, April 1820, p. 382.
[94] *Ibid.*, January 1802, p. 90.

with a great many non-Jewish families in Britain of the highest intellectual calibre.

(5) As stated, Moses and Deborah de Medina also had five daughters. Ester was named after Deborah's mother. She married Solomon, her uncle (her father's brother), one of Joseph alias Aaron de Medina's sons. His full name was Solomon Hiskia alias Francesco. This latter name is additional evidence of Francesco de Medina being the father of both Selomoh de Medina and his brother Joseph alias Aaron.

Solomon Hiskia married Ester on 7 Tamuz 5474 (9 June 1714).[95] He is described in the marriage contract as a 'London Merchant'.[96] The bride's portion was £3,000, out of which £1,500 was to be received from her father and £1,500 from her grandfather, Sir Selomoh.[97] They left for Amsterdam soon after the wedding.[98] Solomon was active there in his father's firm, Joseph Medina & Sons, which he had joined in 1702.[99] While in Holland he became a shareholder in the Utrechtsche Compagnie.[100] In 1725 he was Parnas of the Sephardi synagogue of Maarsen.[101] He returned to London with his wife soon after Sir Selomoh's death,[102] and he brought most of his goods with him.[103] From then onwards, they resided in London permanently.[104] In 1739 he signed the petition for the increase in the number of Jewish brokers in London.[105] They had no family, as, soon after her arrival in Amsterdam in 1714, Ester was taken ill and apparently did not recover. Solomon Hiskia became the final executor of the wills of Selomoh and Ester d'Azevedo, which led to a series of court cases, referred to elsewhere.[106]

(6) Rachel, another daughter of Moses and Deborah de Medina, married Jacob de Aharon Lamago, on 11 Tebet 5478 (4 December 1717).[107] Rachel died between 3 June 1727 and 1 September 1728.[108]

(7) Ribca, the fourth daughter of Moses and Deborah de

[95] *Bevis Marks Records* II, p. 70, No. 202.   [96] Lansdowne 558, f. 69.
[97] *Ibid*.   [98] *Ibid*.
[99] P.R.O. C.11/1570/30.   [100] Zwarts, *op. cit.*, *loc. cit.*
[101] Jacob Zwarts in *Nieuw Israelietisch Weekblad*, Amsterdam, Nov./Dec., 1927.
[102] Lansdowne 558, f. 69.   [103] *Ibid*.   [104] P.R.O. C.11/806/11.
[105] Abrahams, *op. cit.*, p. 93.   [106] See pp. 58–60.
[107] *Bevis Marks Records* II, p. 71, No. 226.
[108] Selomoh's will (Appendix C); P.R.O. C.11/507/2.

APPENDIX B

Medina, died in infancy, and was buried on 7 Tamuz 5472 (11 July 1712) at Mile End, in row 10.[109]

(8) Jochebeth died unmarried.

(9) Abigael married Isaac de Abraham Haim Mendes on 13 Ab 5481 (26 July 1721).[110] They had two sons: Moses (named after Abigael's father) and Abraham (named after Isaac's father).[111]

[109] Burial Register, Beth Holim.
[110] *Bevis Marks Records* II, p. 73, No. 277.
[111] Both referred to in Moses de Medina's will, Isham 61.

APPENDIX C
(a) Wills

THE WILL OF SIR SOLOMON AND LADY (ESTER D'AZEVEDO) DE MEDINA[1]

IN THE NAME OF GOD AMEN on the third day of June in the year one thousand seven hundred and twenty seven about four of the Clock in the Afternoon before me John Sythoff Notary Publick admitted by the Court of Holland residing in the Hague in presence of the Witnesses after named Appeared Sir Salomon de Medina Knight and Madam Ester de Medina D'Azevedo marryed Man and Wife dwelling here in the Hague to me Notary well known he Testator being sick and she Testatrix in indifferent health both having and using their understanding and Memory well as outwardly appeared who not only considering the ffrailty of Human life the Certainty of death and the uncertain Hour thereof, but seeing also that they Testators by the Direction of the Almighty are now become reconciled with their Son in Law and Kinsman Mr. Moses de Medina and likewise with his Brothers Isaac and Salomon alias Ffrancisco de Medina therefore they Testators were determined by mutual last Will to dispose of their temporall Goods which Almighty God hath graciously granted to them in this World But ffirst and principally they Testators recommend their souls unto the infinite Mercy of Almighty God and their Bodyes to a decent Buryall in the following manner viz.

1. concerning which it is their Testators Will and Desire that upon the death of the first deceasing of them both there shall be bought at Ouderkerck on the Amstel by the City of Amsterdam in the buryal Ground there belonging to the Portuguese Jewish Nation two or if necessary three buryal places and that their Testators two Bodyes shall be laid and interred therein but that the said Interments shall not be made before nor the Coffins sooner closed up than till thirty Hours after eithers death.

[1] Sir Solomon's will of 3 June 1727, British Museum, London, Auber, 273, and Isham, 150.

APPENDIX C 95

2. And hereupon proceeding to their intended Dispositions they Testators declared to revoke and annull all their former Dispositions of last Will which they Testators either jointly or severally may have at any time passed or granted under what denomination soever the same may be now thereof excepted but more especially the Testament made by them Testators before me Notary ye. 28th Novem^r 1726.

3. And disposing afresh they Testators declared their Will to be that upon the first Sabbath day after the decease of either of them both there shall be offered up or given to the common Chest of the Congregation of the Portuguese Jewish Nation at Amsterdam called Talmud Torah the Sum of Twenty five Gilders Item on the seventh day to the Officers of the said Church and to the said Common Cash together the sum of ffifty Gilders and then likewise for a particular Legacy the Sum of ffive hundred Gilders as also the Sum of ffifty Gilders on the thirtyeth day and Lastly upon the eleventh Month a like Sum of ffifty Gilders and likewise at the same times there shall be distributed to the Officers and the common cash of the Portuguese Jewish Nation here in the Hague called the Honendal namely upon the first Sabbath day to the Cash of that Congregation twenty Gilders and at each of the three other aforesaid times to the Officers of the said Church and to the said Cash of the Community a sum of Thirty Gilders and on the seventh day likewise as a particular Legacy a Sum of ffive hundred Gilders.

4. Ffurther each of the Testators severally and apart disposing (and first the Testator Salomon de Medina) he Testator declared his Will and Desire to be that the Executors and Administrators hereafter named shall immediately after his decease be obliged to pay to his loving wife Ester de Medina D'Azevedo the contents of her Kthiba amounting with the advance to the Sum of fifteen thousand Gilders Bank Moneys which payment shall be made unto her out of the first and readiest moneys or Effects of his Testators Estate he Testator desiring in like manner that prompt payment shall be made to his said Son in Law Moses de Medina together with his said Brothers Isaac de Medina Senior and Solomon alias Ffrancisco de Medina as representing Joseph de Medina and Sons the Sum of Seventy two thousand five hundred Guilders Bank

money by virtue and upon Account of the mutual Adjustment of Accounts and Agreement which already thereupon hath been passed under hand the 20th. May last past and whereof a Notarial Act is yet to be executed and further (after the decease of both the Testators) a like Sum of Seventy two thousand five hundred Guilders Bank Money also by virtue and in pursuance of the said adjustment and Agreement.

5. Moreover it is his Testators desire that immediately after the decease of the Testators there shall be retained as put out by the Executor or Executors of this Will upon his or their Names or if it should be found necessary—upon the Name of the Portuguese Jewish Church at Amsterdam called Talmud Torah or else upon the Names of Messrs parnassins or Rulers thereof the sum of ffour hundred pounds Sterling capital stock in the Bank of England in order to pay the annual accustomary Dividiends thereof after deduction of Commission and Charges of Correspondency to his Testators Neices Ribca Brandon and Sarah D'Azevedo during their lives viz. one half to the one and the other half to the other respectively during their Lives as aforesaid and after the decease respectively of his said Neices Ribca Brandon and Sarah D'Azevedo the said yearly Dividend shall be payd and employed towards the Subsistance of the Innocent and incapable Son of his Testators Nephew Isaac de Medina Senior named Aron de Medina after whose decease the said Capital of ffour hundred pounds Sterling shall by the said Executor Executors or parnassins or Rulers be delivered over and transferred in property viz. to his Testators Grand Son Salomon de Medina Moses Son the Sum of one hundred and fifty pounds Sterling a like Sum of one hundred and ffifty pounds Sterling to his Brother Abraham de Medina and the remaining Sum of one hundred pounds to his other Brother Isaac de Medina and by decease of one or them all to their respective lawful Descendants or Heirs.

6. Ffurther he Testator bequeaths immediately after his decease to the Portuguese Jewish Church at London called Sahar Asamaim the Sum of ffifty pounds Sterling.

7. Moreover he Testator declared his Will to be that in the Room of the respective Donatives of all his Testators Jewells set or unset which he Testator at several times hath made and hath presented to his dear and loving Wife Ester de Medina

D'Azevedo so as he hath noted the same down in his Books in the Month of January 1726 there shall be left to his said Wife in property the thousand pounds Sterling South Sea Annuityes in England which he Testator hath standing in her Name and also the Sum of ffive thousand Guilders for such Jewells and Plate which she shall please to retain computing the same according to the value or Estimation which shall be found thereof in his Testators Books and besides all which all such Moveables as she shall please to keep for her use in compensation of which Donative by the 15th Article of this Will a donation is likewise made to Mr. Moses de Medina.

8. further he Testator bequeaths (after the decease of both the Testators) to his Neice Simha Brandon married to Aron Arias at present living at London in England the Sum of Two thousand Gilders comprized in an Obligatory Act passed by the Testators to her Use, and which is compensated by this Legacy so that the said Obligatory Act upon payment of the said Legacy shall be delivered up.

9. Also he Testator declared by these presents to bequeath to Mrs. Ribca Aboab ffonseca the Sum of nineteen hundred Guilders viz. ffour hundred Guilders immediately after the decease of the Testator and the remaining ffifteen hundred Guilders after the decease of both the Testators and likewise to his clerk David Roiz Monsanto a like Sum of Nineteen hundred Guilders viz. ffour hundred Guilders immediately after the decease of him Testator and the remaining ffifteen hundred Guilders also after the decease of both the Testators.

10. Item he Testator after the decease of the Survivor of both the Testators bequeaths to the daughter of Abraham and Sarah Mementon (which Sarah Mementon is dead at Surinam) the Sum of two thousand Guilders out of a particular Affection which he the Testator hath had for the said Sarah Mementon which Sum he Testator hath promised by a written Act or missive as well to her father, Abraham Mementon as to her Husband David Usiel Cardoso and also to her Brother Isaac Mementon the sum of ffive hundred Guilders.

11. ffurther he Testator bequeaths immediately after his decease to his good ffriend Mr. Rabby David Nunes-Torres the Sum of ffour hundred Guilders.

12. Item he Testator bequeaths also immediately after his decease to his good ffriend Mr. Ffrans van Limburch two of the silver Candlesticks such as his said Wife shall think fitt to design hereto.

13. Also he Testator immediately after his decease doth bequeath to the Chamber Servant who shall live with him at the time of his death being a Jew the Sum of Two hundred and fifty Guilders but if of another Religion the Sum of one hundred Guilders and to the Chambermaid who at the same time shall be in his Wife's service the Sum of ffifty Guilders.

14. Also he Testator declared that after his said Wife shall have chosen out of all his Books all those which she shall please to keep to bequeath all the remaining immediately after his decease viz. to Mr. Rabby David Nunes Torres all the Hebrew Books and all the others in whatsoever Languages they may be to his Son in Law Moses de Medina.

15. And concerning the Institution of his Testator's Heirs he Testator declared in the first place to nominate and institute his Daughter Deborah de Medina Wife of the said Moses de Medina and by pre decease of his said Daughter her children and further Descendants by Representation in the legitimate portion wherefore he Testator desires that there shall be reckoned unto her that half of his Goods which are mentioned in the fourth article of the Marriage Contract made and passed upon her Intermarriage with her present Husband the said Mr. Moses de Medina on ye. 15th July, 1692 so as the promises are set forth and expressed in the said fourth Article or otherwise all that which upon the decease of him Testator shall be found to belong to his said Daughter or her said Husband as a promised Dower or Marriage Goods pursuant to the said Marriage Contract out of the Estate of him Testator he Testator willing in such Case that to the said promised Marriage Goods and also upon the said legitimate portion there shall be especially reckoned and imputed the Three thousand pounds Sterl. which he Testator in the year 1702 before his Departure out of England delivered into the hands of her said Husband Moses de Medina so as the same will appear by the old Books of the Testator but with Respect of the Two hundred pounds Sterling Capital in the Bank of England which he Testator hath delivered to his said Daughter

APPENDIX C   99

Deborah de Medina (nor the additionals or ffurnishments which he hath thereupon afterwards paid) in his Testators Will that the same shall not be reckoned or imputed nor likewise shall be reckoned or imputed the ffifteen hundred pounds Sterling which he Testator hath contributed towards the portion of his Grand Daughter Ester de Medina who is married to the said Salomon alias ffrancisco de Medina for that the same shall be held as a voluntary Donative or Gift made to his Testators said Daughter and Son in Law which Donatives so farr as necessary are by these presents ratified by he Testator there being likewise by the 7th Article of this Will a Donative made to his Testators said Wife and so consequently all the said respective Donatives cannot be considered to make out any part of his Testators Estate or Inheritance.

16. And with respect to the other equal half part or of all the Residue of his Estate he Testator declared in the first place to institute his said dear Wife Ester de Medina D'Azevedo only in the usufruct during her life and after her decease the Executors of this Will shall thereout pay and give in property the respective Sums following viz. To his Testators Grand Daughter named Ester de Medina the Sum of ffifteen hundred Guilders to his Grand Daughter Rachel Lamago one thousand Guilders to his Grand Daughter Abigael Mendes one thousand Guilders (which thousand Guilders shall be put out in Effects to and for the use and Behoof of her son Abraham Mendes) to his Testators unmarried Grand Daughter Jochebet de Medina the Sum of Three thousand Guilders for a marriage Gift in case she shall happen to marry with the Consent Assent and Approbation of her parents and in the interim the said three thousand Guilders shall remain in the hands and under the Direction of her said ffather Moses de Medina without making good or paying any Interest for the same he Testator further willing that the Usufruct of the remainder of the said half shall be enjoyed by his Testators said Daughter Deborah de Medina and her Husband Moses de Medina during their Lives and after the decease of the said Usufructarists namely Deborah de Medina and Moses de Medina two fifth parts of the said Usufruct shall be enjoyed by his Testators Grand Son Salomon de Medina Moses Son during his life and after his decease

the same shall be enjoyed by his present Wife in case she shall not remarry (the same remaining under the Direction and Management of the Executors and Administrators for the time being) and after their decease he Testator declared to institute in the property of the Capital of the said two fifth parts the lawful Descendants or Representatives of the said Salomon de Medina Moses Son and in the property of the remaining three fifth parts he Testator institutes viz. in one fifth his Testators Grand Son Isaac de Medina and in the two other fifth parts his other Grand Son Abraham de Medina.

17. ffurther he Testator declared to appoint for Executors of this Will and for Guardians over all the Minors therein concerned and also for administrators of his Goods and Estate namely his Testators said loving Wife Ester de Medina D'Azevedo and his said Son in Law Moses de Medina giveing unto them such ample Power and Authority as any ways by Law is due and may be given to Executors and especially also with Power to either of them in case of Absence or other Hindrance to impower another person by procuration in his or her place respectively to represent their respective persons as likewise in case of death of either of them both to appoint and depute anothir provided with like Power and Authority for Executor and Guardian in the place of the deceased and the said Testators Wife in case shall or may desire any Advice or Counsell of strange persons she shall at all times and Occurrences hapening make use of the Advice of his Testators good ffriend Mr. Rabbi David Nunes Torres.

18. ffurther he Testator declared his Will and Desire to be that the said Executors immediately after his Interment at least so soon as any ways possible shall make an exact and distinct State and Inventory of all what shall be found in his Testators whole Estate in his mortuary House together with all the Effects and all the out and home standing Debts and Demands whereof they shall keep the original for the use of the Executorship giveing an authentick Copy thereof to his said dear Wife and by the said Executors at proper times and Opportunities shall be sold and turned into money all what in his Testators Estate shall be found rentless (nothing excepted) and so comprizing thereunder all the Moveables which his Testators said Wife will not keep to put out the produce of

the whole in Stock of the Bank of England or other publick ffunds all which shall be put in the Hands of the said Executors or if they should find it advisable that for greater ffacility the same ought to be done in the Name of one person only in such case the same shall be put upon the Name of his Testators said Wife Ester de Medina D'Azevedo.

19. further it is his Testators desire that his said Executors shall adjust all the Accounts which shall be found standing open in the Books paying out of his Testators Estate to those who shall be truly his Creditors and also getting in and receiving from his Debtors what they are indebted either amicably by process in Law by Composition or otherwise.

20. And She Ester de Medina D'Azevedo now also disposing separately and apart so she Testatrix declared in the first place to institute for her Heir her Daughter Deborah de Medina (and by predecease of her said Daughter her Children and Descendants by Representation) in the legitimate portion due unto them by Law and in the Usufruct of the Remainder of her Estate she Testatrix declared to institute her loving Husband Sr Salomon de Medina during his life and that after his decease out of the Capital there shall be given and divided in full property as follows to Mr. Moses de Medina the Sum of Three thousand Guilders To her Testatrix's Grand Daughter Ester de Medina the sum of One thousand Guilders to her Grand Son Salomon de Medina Moses Son Two thousand Guilders To her Grand Daughter Rachel Lamego ffive hundred Guilders To her Grand Daughter Abigail Mendes ffive hundred Guilders To her unmarried Grand Daughter Jochebet de Medina Two thousand Guilders for a Marriage Gift in case she shall happen to marry with consent Assent and Approbation of her Parents and moreover concerning the same it shall be done and transacted in such wise and manner as the Testator hath ordered with Respect to the Sum of Three thousand Guilders whereunto she hereby referrs herself To Mrs. Ribca Aboab ffonseca the Sum of ffive hundred Guilders and a like Sum of ffive hundred Guilders to the Clerk David Roiz Monsanto.

21. and in the Usufruct of the Remainder she Testatrix declared by these Presents to institute her said Daughter Deborah de Medina and her Husband Moses de Medina

during their lives and after both their decease two fifth parts of the said Usufruct shall be enjoyed by her Testatrixes Grand Son Salomon de Medina Moses Son during his life the same remaining under the Administration and Direction of the Executors and Administrators for the time being and after his decease she Testatrix declared in the property of the Capital of the said two fifth parts to institute his lawful Descendants or Representants and in the property of the remaining three fifth parts she Testatrix doth institute viz. in one fifth part her Testatrixes Grand Son Isaac de Medina and in the remaining two fifth parts her other Grand Son Abraham de Medina.

22. Moreover she Testatrix declared in her Respect to nominate for Guardian Executor and Administrator her said Husband Sr Salomon de Medina and upon his decease she Testatrix declared by these presents thereunto to nominate and appoint in his stead her said Son in Law Moses de Medina with power of Surrogation but in case her said Son in Law should happen to depart this World without having deputed any other person in his place then there shall be appointed in his place such other person or persons as her Testatrix's Grand Sons Salomon and Abraham de Medina shall find advisable to whom the necessary power is thereunto delegated by these presents.

23. As well the Survivor of them both Testators as the first deceasing declared to exclude from their Estate and to excuse from all Administration and Management (saving their Respect and Worships) the Orphan Masters of the Hague and of all other Judicatures & Orphan Chambers where their Testators mortuary House may happen to fall or the Education of their minor Heirs may happen as also more especially all ffriends and Relations not willing that they or any of them shall trouble themselves with their Estate as fully intrusting all the same to their said Guardians and Executors they Testators reserving unto themselves respectively the power and Authority to alter their respective Dispositions to change the Legacyes and to enlarge or annull so as they by time occasionally shall see good either under their hand signing or before a Notary and Witnesses so as they Testators shall think fitt the which theyr will shall be of such fforce as if it were word for word inserted in this their Disposition.

24. All of which they Testators (the same being plainly read out by me Notary) declared for the present to be their last Will and Testament willing and desiring that the same after either of their decease shall have and take Effect either as a Testament Codicill Gift by Reason of death or so as it may best subsist according to Law although that all Solemnities by Law required may not fully have been observed requesting to enjoy the utmost Benefitt. Thus done and passed in the Hague in the House of the Testators in present of Gerrit Hutten and Lambert van Alphen as Witnesses was Signed S. de Medina E. de Medina G. Hutten L. van Alphen & J. Sythoff Not. pub.

IN the Name of God Amen on the first day of September 1728 before me John Sythoff Notary publick admitted by the Court of Holland residing in the Hague in presence of the Witnesses after named appeared Sr. Salomon de Medina Knt. and Madam Esther de Medina D'Azevedo married Man and Wife dwelling here in the Hague to me Notary knowne who declared that having considered and read over again their mutual Testament passed before me Notary and Witnesses on the third of June 1727 to confirm the same by this their Codiciliary Disposition and also to approve and ratifie it in all its points and Articles so far as the said Testament shall not contradict or be opposite to what is hereby changed altered and annulled and accordingly the Appearer disposing alone he declared that Whereas since the date of the said Will there have hapned to dye his Neice Ribca Brandon his innocent Nephew Aron Van Isaac de Medina his ffriend Mr. Rabbi David Nunes Torres as also his Grand Daughter Rachel Lamego and that consequently the respective Legacyes given and bequeathed to the use of the persons in the said Will are come to case so he Appearer hath annulled and revoked by these presents the Legacy or that part of it which by the fifth Article of the said Testament was devised and bequeathed to the use of his Neice Sarah D'Azevedo (in regard she at present is looked upon in a Condition as not to have occasion for the same) it being now his Appearers will that the ffour hundred pounds Sterling Capital Stock in the Bank of England specified in the said ffifth Article as now (instead of being put in the name of the parnassins of the Portuguese Jewish Nation at Amsterdam) shall and must stand and remain in his Appearers Name out of the Capital which at the time of his decease shall be found to be in his Name in the Books of the said Bank and that the Dividend of the said ffour hundred pounds Capital Stock shall be directly received by or for Account of his Appearers said Wife Esther de Medina D'Azevedo (in case she should be the Survivor) to enjoy the same during her life time and that after her decease of the said Capital of ffour hundred pounds Stock there shall goe Two hundred and fifty pounds to and for the use of his Appearers two Grand Sons Isaac and Abraham de Medina upon the same ffoot and manner as is expressed in the said fifth Article but with respect to the remaining one hundred and fifty pounds which is there devised to his Appearers Grand Son Salomon de Medina Moses Son it is his Appearers Will that the said one hundred and fifty pounds shall not be received by him but that the said Capital of one hundred and fifty pounds shall continue to stand in his Appearer's Name and that the Dividend thereof shall be directly received and enjoyed by the said Salomon de Medina Moses Son during his life time and after his decease by his present Wife Rachel de Medina provided she shall not remarry in such wise and after the order as is now at large specified in the 16th Article of the said Will in relation to the two fifth parts of the Inheritance which is there given and devised to the said Salomon de Medina Moses Son Moreover he Testator declared his Will to be that for the better Account of the amount of the said two fifth parts to the Use of the said Salomon de Medina Moses Son mentioned in the said sixteenth Article there shall likewise remain and stand upon the Name of him Appearer Six hundred pounds Sterling Capital Stock in the said Bank of England in order that the Dividends thereof shall in the first place be directly enjoyed by his Appearers said Wife Esther de Medina D'Azevedo (in case she should be the longest Liver) during her life time and that by decease of both the Appearers the Dividend of the said Six hundred pounds shall not goe to and for the Use of their Daughter Deborah de Medina or her Husband Moses de Medina (as by the said 16th Article it was stipulated about the whole Amount of the said two fifth parts) but that the said Dividend of the said Six hundred pounds

APPENDIX C 105

Capital Stock after the decease of both the Appearers shall be directly had and received by his Appearers said Grand Son Salomon de Medina Moses Son during his life or by his present Wife Rachel de Medina in case she shall be the Survivor and shall not happen to remarry so as concerning the same is likewise declared and expressed in the said 16th Article and by decease of them both or after remarriage of his said Widow Rachel de Medina (in case he may have been the first deceasing) the property of this Six Hundred as of the said One hundred and fifty pounds shall goe to the Use of his lawfull Descendants or Representants so and in such wise as expressed by the said Sixteenth Article with respect to the whole Amount of the two fifth parts of the Inheritance devised unto him it being his Testators Will that the said Seven hundred and fifty pounds Capital Stock shall be reckoned and Imputed as a part of that which the said two fifth parts shall amount unto but this Seven hundred and fifty pounds shall in no wise be diminished to prevent that which shall be hereafter explained seeing it is his Testators desire that his said Grand Son shall at the least provisionally have the Dividends of these Sums towards his Subsistance but without the power thereout to discount or pay any out or Acts obligatory or any other Debts which he may have contracted or shall hereafter contract accordingly she Appearer also disposing declared her Will to be that in Compensation for Satisfaction and instead of the Legacy for Two thousand Guilders and for the better Account of the amount of the two fifth parts for the use of her Grand son the said Salomon de Medina Moses Son in the 20th & 21st Articles of the said Testament mentioned after her decease there shall stand upon her Name in the South Sea Company at London ffive hundred pounds Capital Annuityes of the said Company out of the Capital Annuities which at the time of her decease shall be found upon her Name in the Books of the said Company and that the yearly Interest of the said ffive hundred pounds annuityes shall be directly received by or for account of her Husband Sr. Salomon de Medina (in case he should be the longest Liver) to enjoy the same during his life together with all what further is stipulated to his use in the 20th Article of the said Will and after his decease the Interest of the said ffive hundred pounds Annuityes shall not be enjoyed by her

Daughter Deborah de Medina nor by her Husband Moses de Medina (so as was declared by the 21st Article of the said Will) but the same as then shall be received and enjoyed by her said Grand Son Salomon de Medina Moses Son during his life or by his present Wife Rachel de Medina in case she should be the longest Liver and doth not remarry and by the decease of them both or after the Marriage of the said Widow Rachel de Medina (if he should be the first deceasing) the property of this ffive hundred pounds shall go to the use of his lawful Descendants or Representants so and in such Manner as by her Appearer is further disposed and is expressed in the 21st Article of the said Will as well about the Interest as the Property of the said two fifth parts she Appearer moreover willing that in case the said South Sea Annuityes should be paid off that out of the moneys arising therefrom there shall be bought in by the Executors for the time being Stock in the Bank of England and the said ffive hundred pounds Annuityes or the Bank Stock arising therefrom shall in such case serve viz. the Amount of Two hundred pounds Capital in Compensation for Satisfaction and instead of the Legacy of Two thousand Guilders. and the remaining Three hundred pounds in Abatement of the Amount of the two fifth parts of the Inheritance given to her said Grand Son Salomon de Medina Moses Son by the 20th and 21st Articles of the said Will ffurther he Appearer declared that Whereas at two several times he had assisted his said Grand Son Salomon de Medina Moses Son once with ffour hundred pounds Sterling and once with Three hundred pounds Capital South Sea Annuityes and that he is not in Condition to make Restitution to the Appearer for the same he the Appearer doth therefore hereby remit and discharge the same unto him and it is accordingly his Will that the same shall not be required or demanded of him but nevertheless with this express condition that his said Grand Son from now and henceforth shall be obliged sincerely and exactly to perform the solemn promise made to the Appearer of never to play or enter into directly or indirectly any Game or Wagering wherein he within the Space of one Month should win or loose more than the value of an English Crown or three Guilders Hollands money in such wise that if he comes to act contrary to this promise and that he shall be convicted by his ffather or by his Brothers in an honest and a lawful Wise within the Space of one Month of having at one Game or Wager directly or indirectly won or lost more than the value of an English Crown or three Guilders Hollands money it his Appearers Will that in such case there shall be imputed to him as well the ffour hundred pounds Sterling as the three hundred pounds Annuityes and that the value of this Seven hundred and fifty pounds Bank Stock here before stipulated for his use so that after the decease of both the Appearers the said Sum of ffour and three hundred pounds shall come to the Benefit of his two Brothers Isaac and Abraham de Medina viz. one third of the said Seven hundred pounds to the said Isaac and two third parts to the said Abraham de Medina to enjoy the same upon that ffoot and Manner as is declared in the 16th Article of the said Will with Respect to their said parts of the Inheritance and these respective parts in the said Seven hundred pounds are left unto them as an Augmentation of the portions which are given unto them by the said Articles and accordingly also of the same proper Nature and Quality And Whereas by the death of Mr. Rabby Nunes Torres the Legacy is come to cease of the Hebrew Books in the 14th Article of the said Will mentioned therefore he Testator bequeaths the same to his Son in Law Mr. Moses de Medina to keep them together with the other in the said Article mentioned What is aforesaid the Appearers declared to be a part of their last Will the which they desired together with their aforesaid Will should be likewise punctually performed and followed either as a Codicill Gift by Reason of death or so as the same howsoever named may or can best subsist in Law Thus Done and passed in the Hague in the house of the Appearers in the presence of Lambert van Alphen and Thomas Morayqyn jnr. as Witnesses (was signed S. de Medina). E. de Medina De Azevedo. L. van Alphen Thomas Morayquyn junr. and J. Sythoff Notary publick.

(b) LADY (ESTER D'AZEVEDO) DE MEDINA'S CODICIL[2]

IN THE NAME OF THE GOD OF Israel Amen On the twenty ffifth day of March 1731 before me John Sythoff

[2] Isham, 150.

Notary Publick admitted by the Court of Holland Residing at the Hague and in presence of the Witnesses afternamed appeared Mad^m Esther de Medina D'Azevedo Widow of S^r. Salomon de Medina. Deceased dwelling here in the Hague to me Notary known which said Appearer declared that she with very much sorrow hath come to understand that her respected Son in Law M^r. Moses De Medina of London Merchant hath not only for a good while past been in a Continuall weake Condition but is at present in so dangerous disposition of body That she is apprehensive speedily to heare the News of his death and that she appearer finds herself not only in a very advanced age but alsoe in a weake state of health (altho goeing and standing) and therefore she hath Judged Necessary while God Almighty hath been pleased to grant her perfect understanding and memory to make some alterations in respect of her last Will so as she doth by these presents in manner ffollowing

1. But first and principally she appearer declared still to approve and hold for firme and valid all the points of her last Will and Testament so as the same are Contained and Comprised in the Mutuall last Will and Codicill by her Testatrix Jointly with her said Husband deceased both passed before me Notary on the third of June 1727 and the first of September 1728 hereby Confirming the same so farr as they shall not Contradict or be Repugnant to this her further Codicill.

2. And accordingly she appearer declared that by the 21st Article of her said Will she hath Instituted in three ffifth parts of the Residue of her Estate vizt. her Grandson Isaac De Medina in one ffifth part and in the Remaining two ffifth parts her other Grandson Abraham De Medina concerning which she appearer doth now make this alteration that the said three ffifth parts shall be enjoyed by the said Isaac and Abraham De Medina each for a halfe part and so in equall portions.

3. And Whereas she appearer by her Codiciliary disposition hath willed that in compensation for satisfaction and Instead of the Legacy of two thousand Guilders and for the better account of the amount of the two ffifth parts to the use of her Grandson Salomon De Medina Moses Son in the 20th and 21st Articles of the said will mentioned after her decease there should stand in her appearers name in the South Sea Company

at London five hundred pounds Capitall Annuities of the said Company Now it is her Will that Instead of the said five hundred pounds South Sea Annuities there shall remaine standing in her Name and that even upon the same foot and to the use of the same person or persons as mentioned with respect to the five hundred pounds in her said Codicilliary Disposition seven hundred pounds South Sea Annuities

4. And as she appearer by the 20th Article of the said Testament hath willed that there should be given to her unmarryed Grandaughter Jochebet De Medina the sume of two thousand Guilders for a Marriage Gift now it is her Will that there shall alsoe remain standing upon her appearer's Name two hundred pounds South Sea Annuities and that the said two hundred pounds shall be given or Transferred to the said Jochebet De Medina in the room of the said two thousand Guilders at the time she shall happen to marry provided that the same be done with the full Consent and approbation of her parents or by default of them and Especially by default of her ffather then with the Consent assent and approbation of her Appearers Executor or Executors for the time being all the further Conditions which with regard to this Legacy are stipulated in the said will remaining Intire so far as the same are not Contradictory hereunto.

5. And with respect to the Executorship which She appearer hath delegated to her said Son in Law Mr. Moses de Medina alone Now she declared thereunto alsoe to appoint and ordain jointly with her said Son in Law (by reason of his continual weak state of health) her Nephew ffrancisco De Medina otherwise called Salomon Hiskia De Medina and in case of the decease of her said Son in Law to leave the Execution of her respective last Wills to her said Nephew ffrancisco De Medina otherwise called Soloman Hiskia De Medina only to whom she not only giveth the power of Assumption and Surrogation but also all the further power and authority which by the said Will is granted to Executors and Administrators she appearer leaving to her said Nephew in Recompense for the trouble which he shall have herein the sume of one thousand Guilders.

6. What is aforesaid she appearer declared to be a part of her last Will desireing that the same after her decease together with her said other dispositions shall be punctually performed and

ffollowed either as a Codicill gift by reason of death or so as the same however denominated may best subsist in Law.

Thus done and passed in the Hague in the house of the appearer in presence of Gerrit Hutton and Lambert Van Alphen as Witnesses who have alsoe subscribed the Minute together with the Appearer and me Notary which I attest

J. Sythoff Not. Pub.

I WILLIAM DE BOLS Notary Publick dwelling in London duly admitted and sworne do hereby Certify and attest that the Translation out of Low Dutch hereunto annexed Contained in one hundred and seven sides or pages (and purporting to be the last Will and Testament of Sr. Salomon de Medina Kn^t. and his wife Madam Esther de Medina D'Azevedo passed before John Sythoff Notary Publick and witnesses at the Hague the 3rd June 1727 and a Codicill passed by the same Testators before the said Notary and Witnesses at the Hague the 1st Septr. 1728 and alsoe a further Codicill made and passed by the said Esther de Medina D'Azevedo before the said Notary and Witnesses at the Hague the 25th March 1731) is by me ffaithfully made and done Witness my hand in London the ffirst day of June Anno Domini 1731:

William de Bols Notary
Publ^k. A°. 1731.

This Codicill should have been Registred before the above Notarial Act

Translated out of Low Dutch upon a stamp of twelve Stuyvers

No. gs

ON the ffifth day of Aprill 1731 Before me John Sythoff Notary Publick admitted by the Court of Holland Residing at the Hague and in presence of the Witnesses after named appeared Mr. ffrancisco de Medina otherwise called Salomon Hiskia de Medina dwelling at Amsterdam but at present here in the Hague to me Notary known who declared to accept the Charge or Commission comprised in the act hereafter Inserted of the Tenor ffollowing—

APPENDIX C   111

On the ffirst day of Aprill 1731 Before me John Sythoff Notary Publick admitted by the Court of Holland residing in the Hague and in presence of the Witnesses after named appeared Mad^m Esther de Medina D'Azevedo Widow of S^r. Salomon de Medina deceased dwelling here in the Hague to me Notary known who declared that her said deceased Husband by his last Will and Testament passed before me Notary and certain Witnesses on the third day of June 1727 had appointed for Executors of his said Will as Guardians over all the Minors who may be Concerned therein and also for Administrators of his goods and Estate the appearer his Wife together with his son In Law M^r. Moses de Medina Merchant of London in England with power in case of the decease of any one of them both for the survivor to appoint and depute another person provided with like power and authority for Executor and Guardian in the room of the deceased that after the decease of the said Testator the said Executors and Guardians did accept and Exercise the s^d. Commission and Authority to them granted by the said Will that thereupon a few dayes since the said M^r. Moses de Medina hath happened to dye and whereas the Appearer judgeth it Necessary and her self also to be obliged for the benefitt of the Estate which the said S^r. Salomon de Medina dyed possessed off that another person be appointed and deputed in his Room Now therefore making use of the power and authority given and granted unto her by the said Will she declared by these presents to Nominate and Depute for Executor and Guardian in the room of the said Mr. Moses de Medina deceased her Nephew M^r. ffrancisco de Medina otherwise called Solomon Hiskia de Medina dwelling at Amsterdam to the end to take care of the said Guardianship and Executorship with such Authority and power as the said Testator hath given and Granted to his Nominated Executors and Guardians and that hereof it may at all times appear she Testatrix hath thought fitt to pass this present act hereof which is done in the Hague in presence of Jeromino Soarez Da Veiga and David Roiz Monsanto as Witnesses who have subscribed the Minute together with the appearer and me Notary which I attest was signed J. Sythoff Not. Pub: And therefore the Appearer promises to behave himself in the premises as an Executor Guardian and Administrator is obliged and ought to doe pursuant to the Will of the said S$^{o.3}$ de Medina in the aforegoing Act mentioned and to the end the said acceptance may at all times appear he appearer hath thought fitt to pass this present Act which is done in the Hague in the presence of Lambert Van Alphen and Jan de Blieck as Witnesses... who have subscribed the Minute together with the appearer and me Notary which I attest.

J. Sythoff Not. Pub:

ffaithfully Translated by me
Lond$^{n.}$ the 10th May 1731

William de Bols—Notary:
  Publ. A$^{o.}$ 1731:—

[3] 'Sir' in original.

APPENDIX D

Solomon de Medina's Answer to Moses de Medina[1]

To the Most Honble Robt Earl of Oxford and Mortimer Lord High Treasurer of Great Britain

The Answere of Sir Solomon de Medina to Moses de Medina's Memoriall presented to Your Lord^sps 20 June 1712.

My Lord

Finding myselfe sencesibly touched in my Reputation by the wrong suggestions and insinuations contained in the above said Memorial I begg leave to lay the whole matter in a clear light before Your Lord^sps.

First, the said Moses de Medina informes Your Lord^sps that he hath supported me with his money and credit (which I utterly deny) for it was my money and credit that supported him, and to say for 5 yeares last past, when in fact he was interessed, and had the management only for the first 4 years; for in the year 1711 Exclusive he has no wayes concerned.

It is well known how bare he came in to my house and because he was a Branch of my family gave him my Daughter in marriage with a suitable Portion, and when I left England he was worth no more than what he gott in my house, by giving him a Share in my Commissions of the Provedorie wch he owned was worth to him above £2,500 —?—, and was the foundation of his credit he soe much Braggs of.

As to his faithfull Services as he calls them, he has them still to prove the parties concerned refusing to pass his accounts without further satisfaction.

He proceeds to acquaint Y^r Lord^sps that he is ruined in Estate and Reputation by means of £30000 in Bills drawne by him on me, and on my Account that came back Prottested. It is evident non of those bills were drawne on my Account only,

[1] See page 50. (*S.P.Tr.* T.1.154/10A.)

but on account of the Provedorie, wherein he and his Father were concerned a Third with me, who combined together to lett those bills come back Protested to insnare me as the acceptor haveing before furnish[t] him here wth a good part of the money towards discharging of them w[ch] they deviated to their owne private uses and suffered those bills to come back Protested.

As to the £6000 which he says he was forced and compelled to lay down to my £12000 to save his credit it was butt his proportion of his third w[ch] as soon as I heard he was prevailed upon to pay I freely condescended to divert and apply the said £12000 that were ordered on my account, for the year 1711 (in w[ch] he was no wayes concerned) towards paying of this Bills that come back Protested, w[ch] were drawn for Account of the Provedorie for the years 1709 and 1710.

I never had one penny of his money in my hands, so it is proposterous to averr his whole estate lyes in my hands.

I never owned the agreem[t] made the 20 May 1712 and never Declined it (he performing his part of the said Contract) although the £12000 I oblige myself to pay out of that money that will appear Due for the Service of the year 1710 ought to goe in equal proportion wth the former; my £8000 to his £4000 being his third, yet in patternall Kindness I condescended to it to make him Easy.

What he says that by the best interposition he could not persuade me to lay the Accounts clear before Your Lord[sps] I humbly appeal to Your Lord[sp] wither in all the Memorials I have the honour to present to Your Lord[sp] I did not ever complain that I could not have the said Accounts past although they had layen soe long before the Hon[ble] Mr. Bridges and afterwards before him and Mr. Cardonell, upon Your Lord[p] reference to that and, though he falseley alleges, he could never prevail with me to lay them clearly before Your Lord[sp] which being now before the Auditor wth a Ballance due to me thereupon of about £25000 of w[ch], I most humbly beseech Your Lord[sp] to order paym[t] to me of att least £13000 to Answeare those Protests, for which the Creditors are soe very impatient that if not Complyed with, I shall be totally ruined.

His calling the creditors mine only, they are otherwise as much his as mine he being the Drawer, and I, only the acceptor besides his being a third concerned with me as above said.

His laying a demand upon me of £14000 instead of £12000 is to bring in £1900 being my two thirds of £2700 that he says he drew for Lisbon on this Account which are already included in the £12000 as clearly appear by the said Contract.

His desiring that money to be paid into a third hand is plainly against the said Agreem^t, by it that money is expressly to pass throw my hands and endeavouring to putt it out of that Chanell, is to Null the Agreem^t, and putt a disgrace upon me.

Lastly he says "he is ready to refer over accounts to whome I please in order to it" I offered beforehand and since he gave in this Memorial (by Peeter Hennquir Junior) to enter presently in to a Bond of Award wch he refused to do, by all w^{ch}, Your Lord^{sp} may plainly see how well grounded his other Remonstrances are.

My Lord, As I have the honour to be Contractor of Bread and Bread Waggins for five years last past, I have reason to hope that as I am Contractor and have soe well performed the service (to my utter ruine) Your Lord^{sp} will think me able, and worthy to receive and give Discharge for the money Due by the Contract, and that Your Lord^{sp} notwithstanding the ingratitude of my Son-in-Law will rather compationat my case than render it yet harder by making me subject to my Son-in-Law's designs of quashing my Credit to sett up himself, w^{ch}, he has basely done by insinuating in his Memorial and to the Creditors that had he not acquainted them of the last £12000 I issued out, it was to be feared none of them should have been paid, although my interest & safety depends upon paying what is my due, And I hope Your Lord^{sp} and the Publick has not soe ill an opinion of me, as to imagine that I have any other Designe in this matter, but to apply the money due, to the satisfaction of my just debts as Your Lord^{sp} shall see when you are pleased to order immediate payment of said £13000 and the residue when the Auditor shall report what is further Due to me.

I am, wth all Duety and Gratitude

APPENDIX E

Moses de Isaac Dias's Dedication*

DEDICATION
to the
Very Noble and Excellent
Gentleman
Solomon de Medina

Sir,

Whoever were to praise the deeds of Your Honour and repeat your eulogies, to admire your renown and establish your heroic fame, would not find it necessary to crave for inspiration and contrive ideas such as poets use for their fictions, because fiction itself with its enhancement cannot reach the actual truth of such sublime virtue as is in your guileless and generous spirit.

They would say about Your Honour what was said about the wisest of men, the other Solomon, by the Queen of Sheba: "You reached Wisdom beyond Fame." She fell very short in her praise of you.

She could have acted like another Great Queen, Her Britannic Majesty who, putting aside her Sovereignty, frequented your Noble House in spite of jealousies, and in order that you should be well received conferred upon you the title of Citizen among her People and the title of Knight among her Nobles.

But Your Honour has within his breast other purposes, higher considerations than worldly dignities, and is without comparison more at home in the Divine Cult and meditation of the Sacred Law, which from earliest youth led you to be a Citizen of the Divine Fields, a Courtier in the Celestial Abode. I refrain from citing further praises; I shall not refer to your illustrious ancestry, famous progeny of Your Honour, fearing your modesty may reprehend me. Only with the excuse of my daring I must mention your generosity accompanied by ardent zeal, which was the reason for my boldness, a Magnet that attracts me to offer you the efforts of my pen which I place, dedicate and consecrate on the altar of your affection with all due respect, not doubting that you will accept it benignly, whether it be because of your accustomed benevolence or because of the similitude and relation existing between the Comment and the Work, the practice of your Honour and the theory of my pen; and although Your Honour would read more easily in your own person all the virtues which I behold in the Holy Scriptures, so perfectly you imitate them and know how to obey them with your actions, it is appropriate for the virtuous to take pleasure in the contemplation of the virtues. Then will my efforts attain their splendour, having reached their Sphere. My discourses will be clear and intelligible when they are joined to the realistic concepts of Your Honour, for at sight of your examples and acute comparisons, what the pen was unable to explain will be amplified with deeds.

May Your Honour grant in his great generosity that under the wing of such secure protection these efforts attaining repute, defended with the sword of the Law and the fire of their zeal may fly like another Cherub in Paradise, which I attest this work will be in effect, when these leaves are joined to the fragrance and abundance of its virtuous fruits.

May God in His immense Piety keep the illustrious person of Your Honour for long and protracted years with the happiness which the World and this humble Servant wish you.

Your obedient Servant

MOSES DIAS

[* The dedication above is the author's translation from the Spanish. Moses Dias wrote a book, *Meditaciones Sobre la Historia Sagrada de Genesis* . . ., published in Amsterdam in 1697; he added this Dedication to Solomon de Medina in the second and enlarged edition published in Amsterdam in 1705, to indicate his gratitude to Medina, who had supported him.—Ed.]

APPENDIX F

Glossary

Ascama, *pl.* Ascamot – articles of the constitution of the congregation.
Besimantob – 'with good auguring.' The regular formula for congratulating on election to office.
Cautivos – Jews held in captivity.
Codes – *see* Kodes.
Echal – *see* Hechal.
Finta – special tax.
Gabay – Treasurer.
Haham – Ecclesiastical head (chief rabbi) of the congregation.
Hanuca – Festival of Lights; Festival of the Maccabees.
Hatan – Reader; congregational minister who reads and chants the prayers.
Hatan Bereshith – 'Bridegroom of the Beginning.' A member of the synagogue called to the reading of the first portion of the Pentateuch at the Festival of Simchath Torah (Rejoicing of the Law).
Hatan Torah – 'Bridegroom of the Law.' A member of the synagogue called to the reading of the final portion of the Pentateuch at the Festival of Simchath Torah.
Hebra – society attending the sick and burying the dead.
Hechal (Echal) – Ark in which the Scrolls of the Law are kept.
Imposta – congregational dues assessed on income.
Kahal (Kaal) Kados – Holy Congregation.
Kodes or Kodez – sacred, holy.
Mahamad – standing or executive committee of the synagogue.
Miseberah – prayer in honour of a person or institution.
Misva, *pl.* Misvot – act of religious duty.
Parasah – weekly portion of the Pentateuchal cycle of reading.
Parnas, *pl.* Parnassim – Warden.
Pessah – Passover.
Promesas – offerings.
Reby, Rubi – teacher.
Rosh Hashanah (Ros Assana) – New Year Festival.
Sedaca – general funds of the synagogue.
Senhores – gentlemen.
Sepher (Torah), *pl.* Sepharim – Scroll of the Law.
Shabbat Bereshith – The Sabbath on which the first portion of Genesis is recited.
Sochete Bodech – Ritual meat slaughterer and inspector of its fitness for food.
Succot – Festival of Tabernacles.

Talmid, *pl.* Talmidim – student.
Talmud Torah – institution for religious study; also the Sephardi synagogue at Amsterdam.
Teba – platform in synagogue for the Reader.
Terra Santa – Palestine.
Yahid, *pl.* Yehidim – member of the synagogue.

Dr. Oskar K. Rabinowicz

OSKAR K. RABINOWICZ

A BIOGRAPHICAL SKETCH

By Judith K. Tapiero and Theodore K. Rabb

Central Europe in the late nineteenth century was the home of one of the most brilliant cultures in Jewish history, a culture that ranged from great Talmudic scholarship to the first stirrings of the modern Zionist movement. Within this diversity, it was the milieu of traditional learning from which Oskar Rabinowicz ultimately derived. His father, Yehuda Rabinowicz, grew up in the household of the Sadegorah Rebbe, whose spiritual and religious ideals became a powerful and abiding force in the family's life. Yet the tension between the various tendencies within Judaism at the time was also at work, and young Yehuda, both devout and enlightened, eventually found the famous Rabbi's house too confining. He ran away to Vienna, later living in Aspern, Marienbad, Boskowitz, and Brno—a cross-section of the Austro-Hungarian Empire—earning his living mainly as a Hebrew teacher and cantor. Only after World War I did he go into business and enter more fully into secular life.

While in Vienna Yehuda got married and had his first child, Rosa, who was to be a talented pianist. The couple were to have three more children, two boys and a girl, the first of whom, born in 1902, was Oskar. The family was at this time in Aspern, but when Oskar was four years old they moved to the tiny town of Boskowitz, where the remainder of his youth was spent. It was in that compact ghetto, and in the cultured atmosphere of his home, that his interests were shaped which were to remain with him for the rest of his life. A pervasive love of painting and music—he began playing the violin at the age of 5—was born, as was a commitment to Jewish traditions and especially to the Zionist movement, whose first stirrings under Herzl had been centred in nearby Vienna.

The growing passion for Zionism became especially noticeable when Oskar, an excellent student, entered the University of Brno. He stayed only a short time before moving on to the Charles University at Prague, but at both institutions he spent more time in Zionist meetings than in classes. A further distraction was his need to support himself by teaching at the Jewish elementary school of Prague. Indeed, the major event of these years, as far as he was concerned, was unrelated to the university. In 1921, when he was 19 years old, he was invited to attend the Twelfth Zionist Congress at Carlsbad as an aide to Nahum Sokolow, at that time the editor of the *Congress Journal*. The meeting became his initiation into the mainstream of Zionist activity. Sokolow spent long hours with him, discussing the movement, the various party programmes, and Zionist literature. And through Sokolow he met many of the leaders of the day. When, during a speech by Chaim Weizmann, a map fell off the platform, the young man picked it up and received as a reward from the future President of Israel a pat on the cheek—a cheek that he left unwashed for two days thereafter.

The drama and the hopes of Zionism in these heady days following the Balfour Declaration swept him into an enthusiasm for the cause that was to dominate his life for more than twenty years. While continuing his university education, which was to lead to a doctorate in Philosophy in 1924, he also took on his first long-term job as a correspondent for the *Jüdische Volkstimme*, a Brno-based German Jewish paper edited by Max Hickl, a devoted Zionist. His frequent columns, continuing until the demise of the newspaper in 1932, reflected both the variety of Zionist issues in the 1920s and the literary interests he sustained until the end of his life. Reporting on Jewish affairs in general and Zionist Congresses in particular, he not only served as the paper's Prague correspondent but also as the writer of many of its longer, in-depth articles. Two in particular, on Revisionism and on the fight for political Zionism, indicated the direction his own Zionist inclinations were leading.

The blend of journalism and academic pursuits continued

after he moved to Brno in 1924. He had had an opportunity for a post at the University, but it had been made clear that baptism was a prerequisite. The last thing the young activist would have contemplated was conversion to Christianity, and he turned instead to the more open field of secondary education. In Brno he became the co-founder of a new high school, the Jewish Gymnasium, where he also taught history. But at the same time his journalistic and Zionist activities were growing ever broader. During the next decade, he was continuously involved in intensive efforts on behalf of the Zionist movement, as both writer and lecturer. He covered most of the important Jewish or Zionist meetings of the time, either as the representative of a newspaper or on his own initiative; he lectured widely; and between 1921 and 1935 his work appeared in newspapers throughout Czechoslovakia and Germany— *Selbstwehr*, *Židovske Zpravy*, *Tagesbote*, and the *Zeitschrift für die Geschichte der Juden in der Tschechoslowakei*, to name a few.

The particular commitment pursued in this period was political Zionism. He became an adherent of this school of thought after the Twelfth Congress, and was soon an enthusiastic disciple and collaborator of Vladimir Jabotinsky. When, before the Thirteenth Congress of 1923, Jabotinsky resigned from the Zionist Executive, and two years later founded a new opposition party, the Union of Zionist Revisionists, his devoted supporter followed in his footsteps. The headquarters of the new party alternated between Paris and London, and in 1929 Rabinowicz, who had established the Czechoslovak branch of the party, went to London to work for its central executive.

The departure for England marked the end of a stage in his life. During the previous years he had been engaged not only in teaching, journalism, and Zionist propaganda, but also in writing the only non-Zionist books he ever produced. In 1925 there appeared a revised and extended version of his doctoral dissertation, *Spinoza's 'God' in the Light of Jewish Religio-Philosophical Sources*, and two years later he published a *History of the Jews in Aussee* (Moravia). The latter was his last venture into distant history for many years, an account of the rise and fall of a typical small Jewish community. Henceforward his interests were to be concentrated much more directly on his own world and times.

He remained in London for two years, until 1931, working both for the Revisionists and for his father's financial business. But then, late in 1931, another aspect of the family traditions from which he had sprung asserted itself. His father had long felt that the young man had the capacity to qualify for the Rabbinate; and so, acceding to those wishes, he left for Berlin to study at the Hochschule für die Wissenschaft des Judentums. At this seminary he began two years of preparation for ordination, without, however, neglecting the interests he had developed over the preceding decade. He continued to lecture extensively on Zionism within Germany and elsewhere, travelling through Austria, Rumania, Yugoslavia, and Czechoslovakia. In January 1933 he finally attained the desired ordination, sent his father the diploma as proof, and then immediately abandoned his new profession—the only evidences of his status as a rabbi thereafter were sermons he gave on the High Holy Days at the Pinkas Synagogue in Prague.

His immediate employment upon leaving the seminary could hardly have been more incongruous—in the army. Enlisting for military service, he was made chaplain to the troops. But he was hardly the ideal soldier. He insisted on kosher food, took Jewish holidays off to join his family, and spent much of the time allotted to manœuvres in the synagogue. By the end of 1933 the army decided to release him before he had served a full term, and he quickly resumed his earlier activities.

During the little more than five years that he was still to live in Czechoslovakia his work in Zionism, journalism, and Jewish causes entered its most intensive phase. His place within the Zionist movement had long been clear, firmly on the side of Jabotinsky and his political aspirations. On the other side stood Weizmann, who was convinced that a political solution, the creation of a separate Jewish State or a State with a Jewish majority, was unnecessary. At a meeting of the World Congress of Revisionists in Prague in 1930, which drew delegates from all over Europe, general plans and the growing rift with Weizmann had been discussed, but open confrontation had been avoided. In fact, following his participation in the Congress Rabinowicz went on to serve as a delegate at the Seventeenth and Eighteenth Zionist Congresses in 1931 and 1933. But the conflict was coming to a head, exacerbated by the Revisionists' dismay at the growing control over Zionism that was being wielded by non-Zionists in the Jewish Agency. It was this, as he later put it, that 'made it impossible for true Zionists to remain within the Zionist Organization',[1] and led to the split of 1933. The adherents of Weizmann stayed in the Zionist Organisation and called themselves the Jewish State Party. The secessionists, led by Jabotinsky, formed the New Zionist Organisation, on whose executive Rabinowicz sat from 1935 to 1938. He was also president of the Czechoslovak branch of the new party.

Although most of his energies were now thrown into the Revisionist movement, he also devoted himself to broader Zionist work. He was a member of the Zionist Actions Committee of the Jewish Agency between 1931 and 1935, and he also took over the chairmanship of the Committee to Boycott Nazi Germany in 1934. Moreover, during the three years before the European war broke out he took charge of the machinery set up to provide illegal immigration to Palestine for Jews who had sought refuge in Czechoslovakia from Germany and Austria. Some three thousand Jews were able to reach Palestine during this period as a result of his efforts.

For all these undertakings, his writings were as voluminous as ever. He produced three books in the 1930s. The first, published in 1932, celebrated the 15th anniversary of the founding of the Jewish Legion during World War I. It was a collection of essays on the history and development of the Legion, with an introductory survey of the Legion's fifteen years written by Rabinowicz himself. The theme of the essays—admirably serving the aims of the more militant Zionists—was that the Legion's fighting prowess had won new respect for Jews the world over.

His next book, *When Nations Awake*, again reflected his Revisionist concerns. Published in 1934 and dedicated to Jabotinsky, it explored the elements of Czech nationalism which so recently had made possible the creation of a Czech State. Naturally, each element was shown to have a parallel in the rising nationalism of the Jews—yet further justification for the Revisionists' aims.

[1] Lecture on 'Revisionism' to the Oxford University Jewish Society, 7 June 1942.

The third of these books, published in 1937, had a more general purpose. Entitled *Introduction to the Problems of Ritual Slaughter*, it was intended as a response to a debate that had taken place in the Czech Parliament. At issue was the Jewish custom of shechita, the prescribed method for killing animals so that they met the requirements for kosher food. The Parliament had considered condemning the ritual as inhumane, and his defence not only outlined the traditional laws of shechita, but also assembled evidence to show that it was the most humane possible form of slaughter. He was once again acting, with every means at his disposal, as the defender of his people.

But his most sustained enterprise at this time was in journalism. In 1934 he founded his own newspaper, a weekly called *Der Judenstaat—Medina Ivrit*, shortened two years later to *Medina Ivrit*, whose editor he remained from its first issue, on 7 December 1934, until its dissolution in 1939. Its frankly Zionist name was echoed by its contents, which were devoted to the movement in general and to Revisionism in particular. Beyond its advocacy of Revisionism and the Herzlian concept of a Jewish State, however, *Medina Ivrit* also gave its attention to general issues facing the Jewish community: Jewish culture, education, youth, and a wide variety of social and political problems. Inevitably, one of its major purposes soon came to be the denunciation of Nazism. But its editor also gave its columns over to more scholarly and extended discussions. He himself wrote many of these features, notably a series of articles on Maimonides' vision of the Jewish State, in which he emphasised how closely the medieval philosopher's proposals for the establishment of the State prefigured the Revisionists' own programme.

Appropriately enough, it was as a direct result of his multi-sided public engagements that a major change also occurred in Rabinowicz's private life in the 1930s. In October 1933 he came to the town of Teplice-Šanov, north-west of Prague, to give yet another lecture on Revisionism. Party rivalries were as intense in this small community as they were within Zionism at large, and a rival faction, a leftist youth movement known as Blau-Weiss (Blue-White), sent an undercover agent to this enemy meeting. The spy's mission was to note if any member of Blau-Weiss had dared to attend such a nefarious gathering, but instead found both speaker and speech of far greater interest, and went on to a party in the speaker's honour. Five months later the hapless informer, a sixteen-year-old girl named Rosa Oliner, married Oskar in a ceremony boycotted by her friends because of her great betrayal! The couple moved to Prague, where three years later their son Theodore was born. It was a busy time, with the newspaper, Revisionism, and the organisation of Jewish emigration requiring enormous energy. But all of these activities were about to come to an end in face of the assault on central European Jewry that was gaining momentum across the border in Germany.

Rabinowicz did not restrict himself to denunciations of Nazism during these years. When Hitler invaded the Sudetenland in September 1938, there was a call for general mobilisation throughout Czechoslovakia, and he again entered the army. He was released at the end of February 1939, as the threat clearly became one that the Czechs could not handle on their own. A few days later he received a telephone call from Frantisek Chvalkovsky, the Foreign Secretary of Czechoslovakia, warning him that his name was on the lists the Gestapo had prepared of people to arrest when they entered Prague. His chairmanship of the Committee to Boycott Nazi Germany and his outspoken anti-Nazism made him an obvious target, and Chvalkovsky advised him to be ready to leave at any time. The expectation was that the Germans would invade on 15 April, the anniversary of their invasion of Austria, but instead they crossed the border on 15 March 1939.

He was caught unprepared, but not beyond a last-minute escape. Hearing on the 6 a.m. newscast that the invasion had begun, he went straight to the station and caught the last train to the German border at Eger (Cheb). Then, fluent in German, and taking advantage of the chaos caused by the troop movements, he slipped across the frontier, confident that this was the last place he would be sought and that officials would be too harried to check his documents. His transportation was a peasant, with horse and cart, who was not even stopped at the border. Hitchhiking the rest of the way, he crossed Germany to Belgium, and from there went on to Paris and London. His first income in exile came from a story he wrote for the *Pariser Tagesblatt* describing his escape.

Although he had left both wife and child in Prague, the upheavals of the time were sufficient to keep them from positive identification by the Gestapo. They managed to send most of the family's belongings to London, and in May followed him to the sanctuary offered by England. His sister Martha, miraculously surviving the war, found her way to Paris and later to New York. His brother Kurt was to make his way to Palestine later in the war, but his parents and sister Rosa were to disappear without trace during the German occupation. He never set foot in his homeland again—indeed, when it fell under another foreign influence after the war, he gave up his Czechoslovak citizenship and remained Stateless until he qualified for British nationality in the 1950s. An era had ended, and now, as he approached his forties, his life was to take yet another turn.

For a number of years he had been helping his father in his business activities, and it was with this experience to guide him that he now sought new ways of supporting his family. With what capital he had, he purchased first a finance company, Clements Trading Company, and after a few years the Anglo-Federal Banking Corporation, an almost defunct enterprise that he revived. Since England, too, was involved in the war, every business played its part in the national effort, and he became involved with a munitions factory that produced parts for guns.

As the German air offensive against London mounted, between late 1939 and 1941, the family moved to Oxford. Here it found itself in the war-time Jewish 'capital' of England, surrounded by other refugees in an almost Continental atmosphere. Here, too, long-time friendships were begun, notably with his fellow-refugee Bela Horovitz, the founder of Vienna's Phaidon Press, and with the Oxford don Cecil Roth, the foremost Jewish historian in the English-speaking world. But sometimes special circumstances were necessary before the foreigners received an open welcome. The postman, for example, frustrated by the strange and overlong names he now had to distinguish, frequently gave up the struggle, and threw all such mail, undifferentiated, on the doorstep. But at this very time Rabinowicz was beginning to explore a new interest—the career of the man he considered the greatest contemporary non-Jewish friend of the Jews, Winston Churchill. In the course of his research, he entered into correspondence with Churchill's private secretary, and when the postman saw a letter bearing the return address of 10 Downing Street, his attitude was instantly transformed. Henceforth he rang the bell and always delivered the mail in person.

Once back in London in 1941, Rabinowicz resumed his public work for Jewish causes, but with less emphasis on political activism. Instead, literature, scholarship, and long-range restoration became his main concern. Together with Josef Fraenkel and Joseph Leftwich he established the Committee for Zionist Research (1941), whose aim was to compile an accurate account of Zionist history. As first steps, he and his colleagues published studies of Herzl and other Zionist leaders. Two years later he joined the Committee on the Restoration of Continental Jewish Museums, Libraries, and Archives, which was formed (1943) at a conference called by the Jewish Historical Society of England. This meeting brought together representatives of the British Jewish community and of the various communities under Nazi control on the Continent, and resulted in the appointment of Cecil Roth as President and Rabinowicz as Honorary Secretary. Their mandate was to prepare statistical and other material which would be needed after the end of the war to save whatever Jewish cultural materials had survived, but since only the authorities in the American zone gave them permission to search after 1945, they joined in association with the Jewish Cultural Reconstruction of New York. The united group finally set up a joint office in Wiesbaden in Germany.[2] As usual, however, Rabinowicz did not content himself with a few such activities. He was also a member of the Relief Committee for Jews from Czechoslovakia, was elected to the Council of the Jewish Historical Society of England, and served on the Council of Christians and Jews from 1940 onwards.

A major reason for the lessening of his political involvements was the scattering of Jewish leadership by the war, and, more directly, the death of Jabotinsky in 1940. It was also clear that

[2] See Souvenir Brochure, Leeds Home for Aged Jews and Home of Rest, 1950.

the organisations familiar in Czechoslovakia could not be transplanted to England, although he did take part in the reunification of the New Zionist Organisation with the Jewish State Party in 1946. He always regarded political action as an instrument for the moment, not as a long-range avocation, and he now devoted himself to broader Jewish and Zionist aims. His swan song as a Revisionist was a short book published in 1946 bearing the title *Vladimir Jabotinsky's Conception of a Nation*. It was a subject with which he was closely familiar, and whose various dimensions he had long explored. But now his interests were more scholarly then partisan; and, for all the affection he had for his subject, his prime objective was to present Jabotinsky's ideas on modern European nationalism and to analyse those ideas, not to engage in propaganda.

The remainder of the war was taken up by these business, cultural, social, and literary occupations, sustained despite war-time conditions and a major move. For shortly after the birth of a daughter, Judith, in 1943, the family settled in Harrogate, in Yorkshire, to escape the flying bombs, and only returned to London in 1945. It was in this northern town that another incident occurred which revealed the peculiarities of life as a refugee. A director of Clements Trading, Lord Semple, a leading British Catholic, was escorting a Papal Nuncio to the consecration of a new church in nearby York, and he suggested a short stop at the house in Harrogate. The only way the neighbours could explain the visit of the Pope's representative to a foreign Jew's house was by assuming that the entire family had decided to convert to Catholicism.

Upon returning to London in 1945, Rabinowicz took up residence in Hampstead for the next eleven years and entered yet another phase of a career that had already seen so many changes. His principal work was as Chairman and Managing Director of the Anglo-Federal Banking Corporation, through which he came into contact with leading British industrialists, notably Sir Isaac Wolfson. Yet even these undertakings were turned to Jewish purposes, particularly after the establishment of Israel in 1948. A frequent visitor to the house in Greenaway Gardens was Pinchas Sapir, seeking both advice and help in negotiating loans for the new State, and meeting sympathetic leaders of English finance. Local causes also received support.

At one dinner party, when the discussion turned on the needs of a Hillel House for students in London, Wolfson took out his pen, wrote a promise for £1,000 and then went round the table marking down nine friends for another £1,000 each. Thus the first £10,000 was raised to send the new institution off to an excellent start. Other parties aided the Hebrew University, the Midrashia, and similar cultural causes.

The other notable feature of the house was its library of some 16,000 volumes, specialising in Jewish and Zionist history, which had been accumulated in the course of years of collecting. This was always open to serious scholars, and thus, in addition to the businessmen, the house often welcomed struggling young writers and researchers. Rabinowicz always took a personal interest in their work, encouraging them and often seeking out himself the materials they needed. Started by his father, the library eventually became one of the largest private repositories of Jewish and Zionist sources in the world. One of its unique features was a collection dealing with Hebrew grammar, assembled by Yehuda during his travels through Europe, and used as the basis for a book on linguistic Biblical research. To this had been added not only extensive Zionist holdings, but also the letters of prominent Jewish authors and scholars, such as Bialik and Ish Shalom, and rare books uncovered at auctions and bookshops. As a result, the home in Greenaway Gardens reflected the blend of business activities, literature, scholarship, and social concern that was visible throughout its owner's life.

The public activities during these years were, as usual, multi-sided. The committees he served on touched on almost every aspect of Anglo-Jewish life, from the Friends of the Hebrew University, Jews' College, and the British Friends of the Midrashia, of which he was Vice-Chairman, to the Association of Jewish Journalists and Authors, the Memorial Council, and the Jewish Record Office. In addition, he was a founder and governor of England's first Jewish public school, Carmel College, and a member of the London Committee of the Centre of Jewish Documentation, a group which included Professor Max Beloff, Sir Isaiah Berlin, Sir Israel Brodie, Dayan Lazarus, Joseph Leftwich, Dr. Machover, and Cecil Roth. His most notable service, however, was in 1956, when he was appointed, jointly with Lord Rothschild, as co-Treasurer of the Tercentenary celebrations commemorating the return of the Jews to England under Oliver Cromwell. The honour was particularly significant in that the Council of which he was named an officer consisted primarily of men whose families had been in Britain for generations, yet he himself was an immigrant who had arrived only 16 years before.

But his older interests persisted too. He wrote frequently for the *Jewish Chronicle*, both as a reviewer of books on modern Jewish and Zionist subjects and as the contributor of feature articles on such topics as Churchill, Weizmann, and the holdings of his library. Nor did he hesitate to engage in hearty debates in the 'Letters to the Editor' columns on subjects as varied as Ben Gurion's view of history and Chamberlain and Zionism. Not content with reviving his love of journalism, and writing for a number of English journals and newspapers, he also resumed his old inclination for public speaking. In 1954 he delivered the address at the official consecration of the Wanstead and Woodford Synagogue, in return for which he received the key to the Synagogue. And, when asked to give a talk at the annual meeting of the Jewish Historical Society, he picked as his subject a new research interest, Sir Solomon de Medina. Medina, a close friend of the first Duke of Marlborough and the provisioner of England's armies during the wars with Louis XIV, was the first Jew ever to be knighted. The Jewish link with Churchill's ancestor, and Medina's prominence in Anglo-Jewish history, made this an especially appealing subject. He continued his researches after the talk was delivered, expanding his treatment until he completed the present full-scale biography, which remained hidden among his papers until his death.

The passion for writing always remained his first love, and the London years witnessed the publication of a series of books on his lifelong preoccupation, Zionism. The first, written in 1950, was *Fifty Years of Zionism: A Historical Analysis of Dr. Weizmann's 'Trial and Error'*. He had bought Weizmann's book to read on the plane trip back to London after a holiday abroad, and was appalled by what he saw. Going through it as quickly as he could, he immediately began making notes on the distortions of Zionist history that he found. Two weeks later the first draft of the book was finished, written 'to tell my grandchildren the true story—to set the record straight'.

The chief injustices of *Trial and Error* were directed at Theodor Herzl. Weizmann minimised the contributions of the founder of modern Zionism and distorted his ideas, reasserting instead his own, long-discredited criticism of political undertakings. Where Herzl had insisted on the political recognition of Jewish aspirations by the Great Powers, Weizmann had relied on economic growth and settlement, and had denounced all who disagreed. When he used his autobiography to reopen the issue in one-sided fashion, Rabinowicz showed that without Herzl's inspiration, without political recognition, neither the Balfour Declaration nor the establishment of the State of Israel would have been possible. *Fifty Years of Zionism* also revealed the true reasons behind Weizmann's resignation from the Presidency of the Zionist Organisation in 1946 and the real story behind the Brandeis–Weizmann feud, both of which had been glossed over. And the consistent errors and underestimations in the treatment of Zionist leaders were revealed for what they were; as a result, Moses Gaster, Leopold Greenberg, Jabotinsky, Max Nordau, Nahum Sokolow, Menahem Ussishkin, and David Wolffsohn were restored to their rightful prominence in the history of Zionism.

Although Weizmann's adherents never responded to the book, they did all they could to reduce its impact. The Hebrew edition and both English editions were rapidly declared 'out of print' when they were issued in Israel. And when the French translator of *Trial and Error*,[3] Vivian Maspetiol, saw fit to add a résumé of *Fifty Years of Zionism* as an appendix, this version, too, quickly became 'out of print' in Israel. Not until 1967 was Rabinowicz able to find a copy of this last publication in a dusty corner of a Jerusalem bookshop, having searched for it in vain for ten years.

But most of his scholarship at this time was devoted to the early years of Zionism. His first essay on Herzl, a subject to which he was to return repeatedly, appeared in 1951— 'Herzl and England'. This initial study outlined the negotiations which had led to the East Africa Scheme of 1902 and reprinted the relevant documents from the Foreign Office files.

[3] The French title was *Naissance d'Israël*; it appeared in 1957.

In the following year, exploring these materials further, he wrote more extensively on the importance of the scheme. This article, entitled 'New Light on the East Africa Scheme', revealed that the plan and the accompanying negotiations had been instrumental in establishing the close collaboration between the Zionists and the British Government that was to culminate in the Balfour Declaration in 1917.

The last major work Rabinowicz undertook in England was an act of homage to the man he considered the Jews' best non-Jewish friend, the leader of the country that had saved so many Jews, Winston Churchill. He had been gathering material for the book since 1940, and had been received at 10 Downing Street during the early 1940s by Churchill's Personal Secretary, who urged him to delay publication until after the war because of the nature of the materials he intended to use. Although he lectured frequently on the subject, to such audiences as the Hampstead Literary and Debating Society and the First B'nai B'rith Lodge of England, and wrote an article in honour of Churchill's 80th birthday, he did not finally publish *Winston Churchill on Jewish Problems: A Half-Century Survey* until 1956.

In the book Churchill emerges as a constant advocate of Jewish rights and a Jewish homeland. Many of his stands in Parliament indicated this concern: his opposition to the Aliens Bill of 1904; his opposition to the Passfield White Paper of 1930, which limited the scope of the National Home; and his opposition to the White Paper of 1939, which abandoned the concept of a Jewish National Home and aimed at stopping immigration into Palestine. Repeatedly his policies were based on a firm belief in Jewry and its aspirations: 'The Jews of the world . . can look to Sir Winston as a good friend and one who has always had the interests of the Jewish people at heart.'[4]

Rabinowicz's intention was to follow this volume with a sequel devoted to Churchill's stands on Zionism and the creation of a Jewish State. In the late 1950s, while gathering material for this project, he met Eleanor Roosevelt and received her permission to do research in the Roosevelt Archives in New Hyde Park, N.Y., because between 1939 and 1945 the two

[4] *Jewish Chronicle*, 26 November 1954, p. 19.

leaders had corresponded frequently on the Jewish question. These documents were vital to the book and he was allowed to microfilm extensively—on condition, however, that none of the sources be published, since they were still classified. That second volume, though complete and the basis for others of his writings, has yet to appear.

The last phase of this many-sided career opened in 1956, when the family moved to the United States. Rabinowicz retired from the bank and now determined to devote his time to his writing. Although he had left the business world, he kept up his many English friendships while forming new ties from his Scarsdale, New York, home. He became involved, again, in a whole series of cultural and social organisations with extensive connections throughout Jewish life, all the while giving his literary and scholarly interests the sustained attention they had never previously received. In his last years he was even able to indulge once more the love of painting that he had first developed in his youth.

The public commitments were soon as numerous as they had been in London. Rabinowicz served on the Board and on the Library Committee of the Jewish Theological Seminary, on the Editorial Board of the Jewish Publication Society, and on the Editorial Board of the *American Zionist*. In addition, he joined the American Friends of the Hebrew University, was elected to the Council of the American Israel Cultural Foundation and the Executive Board of the World Jewish Congress, and became Vice-President of the Conference on Jewish Social Studies. Returning to the work of his youth, he helped found the Society for the History of Czechoslovak Jews and acted as Editor-in-Chief for the Society. At the same time he joined the Advisory Board to the Judaica Department of Brandeis University, and became both a contributor and the Departmental Editor for the History of Czechoslovak Jews for the *Encyclopedia Judaica*. Yet during these last thirteen years of his life he also managed to write eleven articles and books, not to mention contributions to various newspapers and encyclopedias. The prolific output continued until the very end.

His first American book appeared in 1958: *Herzl, Architect of the Balfour Declaration*, a work that had been in preparation for a number of years. Here Herzl was revealed as the man who had originally introduced England's political leaders to the principles of Zionism. He had met them during the negotiations over the East Africa Scheme in 1902, had communicated to them the ideals that were beginning to move European Jewry, and had thus paved the way for later efforts. When Sokolow, Weizmann, and others started to work towards the Balfour Declaration of 1917, they found their way eased by the familiarity with Jewish problems that the English had gained from Herzl. No stronger case could have been made against the slighting of Herzl's importance in Weizmann's autobiography, and the book can be seen as the conclusive successor to *Fifty Years of Zionism*.

He returned to Herzl two years later, in an article entitled 'Herzl and England' which explored the pioneer Zionist's real aims in suggesting East Africa as a possible home for the Jews. It turned out that he was not abandoning the hopes for Palestine, but executing a subtle political manœuvre. The very negotiations represented a recognition and confirmation of the legitimate demands of Jewish nationalism and thus formed a crucial stepping-stone towards the Balfour Declaration. In the same year Rabinowicz published his last article on his great hero. 'Herzl the Playwright', however, followed a rather different theme. Gone was the Zionist polemic; instead, there was a warm appreciation of an often neglected aspect of Herzl's versatile career during his early years as a Viennese intellectual. Through his 28 plays and sketches had run two themes—the importance of the family as the essential unit of society, and the evil and corrupting influence of money on family bonds. 'What emerged was the wish for a new kind of society, where family life was upheld and the power of money destroyed'—a society that he later foresaw as coming into existence as a Jewish State.[5]

Digging even further back into the origins of Zionism, the next book, *A Jewish Cyprus Project: Davis Trietsch's Colonization Scheme* (1962), examined one of the many schemes that had flourished in the 1890s and early 1900s. Trietsch's hopes for a

[5] Theodore K. Rabb, 'The Œuvre of Dr. Oskar K. Rabinowicz', in *The Jews of Czechoslovakia: Historical Studies and Surveys*, Vol. II, Philadelphia, 1971, p. 11. This article provides complete references to all the publications and further discussion of their contents.

Jewish settlement in Cyprus which would then spread throughout the Middle East gained no response either from the British Government or from the Zionist movement, and they soon faded into oblivion. But the contrast with Herzl was revealing; as Rabinowicz pointed out, Trietsch failed so miserably precisely because he ignored the importance of Jewish nationalism. Yet again the political dimension of Zionism was revealed as the critical element in the movement's history.

During these years he had also been concerned with more contemporary issues. Remembering his work for Jewish refugees in the 1930s, he turned his attention to the painful problem of the Arabs displaced by the creation of the State of Israel. He knew that a solution would have to be found, but first the refugees had to be identified and their numbers accurately calculated. To this end he published in 1959 'The Jews and the Arab Refugees', a detailed analysis of the most reliable statistical evidence, which showed that only 371,500, not a million, Arabs had fled their homes. His recommendation was that Israel should take the initiative and declare its willingness to take these Arabs back.

There was time, too, for less weighty investigations. In 1958 he was asked by his synagogue, Temple Israel, of White Plains, to edit the volume celebrating its fiftieth anniversary. The history might be of a different order, but he gave it the same dedication he bestowed on all his projects. A regular synagogue-goer, he worked conscientiously for the congregation, and took considerable pleasure from the handsome volume that resulted.

During the last six years of his life he fulfilled another long-standing dream—to have his own residence in Israel. He had first visited Palestine in 1925, for the opening of the Hebrew University on Mount Scopus; he had returned on his honeymoon in 1934; and he had then been in the Holy Land every year, with the exception of the war years. In 1963 he finally bought a flat on Jerusalem's Ben Maimon Street, where he spent three or four months annually thereafter. These stays brought him a contentment that was the culmination of his life's work. He knew Hebrew fluently, though it was a classical, literary Hebrew rarely heard on the streets of Jerusalem. And he gained a sense of belonging, of having participated with his people in the realisation of ancient hopes, that made all the struggles at last seem worth while.

In Israel, as in England and America, he soon became integrated into local life. His circle of acquaintances in Jerusalem included professors, lawyers, judges, members of the Government, and artists, many of whom he had known in earlier years in Europe. Taking part in public life there as elsewhere, he joined the Board of Directors of Rassco, became an official Founder of the Hebrew University, and frequented the university's activities. He was also a member of the Jewish Historical Society of Israel, a life member of the Weizmann Institute, and one of the Friends of the Haifa Technion. In recognition of his many scholarly achievements, he was awarded the Landau Prize, given to Jewish writers who have excelled in their field.

Perhaps the best testimony to his easy integration into Israeli life was provided by an experience shortly after he moved into the Ben Maimon flat. He bore a remarkable resemblance to Ben Gurion, being of approximately the same height and with a shock of white hair surrounding a prominent forehead. One evening he was standing on the terrace of the flat, which overlooked the Prime Minister's official residence across the street. It was twilight, his face was no longer distinct, and the guards across the street later told him that they had mistaken him for the former Premier, reconnoitring the official residence and making plans for its reoccupation.

His last publications demonstrated again the breadth of the interests he had followed for nearly 50 years of literary activity. In 1965 he brought to an end a controversy in which he had taken part for a decade, and which concentrated on an area of Jewish scholarship totally removed from his previous work. At issue was the forgery of a scroll of Deuteronomy which had been offered to the British Museum by a Mr. Shapira for one million pounds in 1883. The forgery had been exposed at the time, but in 1956, in the wake of the discovery of the Dead Sea Scrolls, Professor Mansoor, of the University of Wisconsin, reopened the case. Drawing on the techniques of Biblical documentary analysis as well as the accounts on the 1880s, Rabinowicz responded in an article, 'The Shapira Forgery Mystery', which upheld the judgment of the British Museum.

When Mansoor debated his findings, Rabinowicz replied with two further articles in 1957, and then summed up the entire episode in 1965 with an article entitled 'The Shapira Scroll: A Nineteenth Century Forgery'.

In the following year he launched into a different kind of research. The Jewish Historical Society of England was honouring his old friend Cecil Roth on his 70th birthday with a book, *Remember the Days*, and his contribution was to compile a complete Roth bibliography, a list that eventually ran to over five hundred items, some of which even Roth had forgotten. Another milestone was celebrated the next year, the 75th anniversary of the *Jewish Quarterly Review*. His article for the commemorative volume, 'Jacob Frank in Brno', took him into an eighteenth-century subject for which he had done documentary research many years before in Czechoslovakia. He traced Frank's life in Brno during the vital years between 1773 and 1784 when he had discovered himself to be the reincarnation of Shabbatai Zvi, had launched a messianic movement, and had gone on to a fate similar to Zvi's, as a convert to Catholicism.

Closer to most of his long-standing work were two articles published in 1967 and 1970. 'Zionist-British Negotiations', in *Essays Presented to Chief Rabbi Israel Brodie on the Occasion of his 70th Birthday*, continued the story of Herzl's negotiations with British leaders. It told of the discussions Herzl's adherents entered into shortly after his death over the possibilities for a Jewish settlement in the Sinai Peninsula. In this way the continuity of contact with the British Government, begun by Herzl, was maintained. Another favourite subject was 'Churchill and Israel', on which Rabinowicz gave a talk in London in 1968 at a meeting celebrating the establishment of a Chair in International Relations at Bar Ilan University. Although only a few months from death, he delivered the talk with the energy that his many friends in the audience immediately recognised as his most characteristic trait. Later published in Vol. XXII of the *Transactions of the Jewish Historical Society of England*, the article summarizes the contents of the second volume on Churchill, emphasising the great Englishman's unflinching support for the Zionist cause.

His last two projects linked Judaism with the two countries

in which most of his life had been spent, and he finished both on the very eve of his death. A book on Toynbee, which is entitled *Arnold Toynbee on Judaism and Zionism*, to be published soon, analysed the transformation of the historian's thoughts about Judaism. Until the end of World War I, when he held an influential position in the Foreign Office, Toynbee had been pro-Zionist. But by the 1930s he had reversed his position and used the Jews as a target for attack throughout his massive history of the world. Rabinowicz had outlined the friendly phase in an article, 'Toynbee's Pro-Zionism in World War I', published in 1968; the book, by contrast, analysed the onslaught and rebuffed it in detail.

The other project was more congenial, because in 1968 Czechoslovak Jews celebrated the 1,000th anniversary of the founding of their community. Lectures, articles, and dinners marked the millennium, which was to culminate in festivities in Prague. The Czech Government was making special arrangements, and he was about to take up an invitation to return to his homeland for the occasion when an invasion again intervened—this time the entry of the Russians into Czechoslovakia. Instead, therefore, the Jews had to remain at home for their celebrations. In November they sponsored a programme at City Hall in New York called 'The Millennium of Jewish Cultural Life on the Soil of Czechoslovakia' at which Rabinowicz, fittingly, gave his last public address. Two months later, however, in January 1969, he took part in one final commemoration, a programme on NBC television in which he, together with Metropolitan Opera conductor Jan Behr, reminisced about Czech Jewish history.

These specific activities were linked to a more long-range enterprise. The Society for the History of Czechoslovak Jews, of which he was Editor-in-Chief, produced its first volume on the *Jews of Czechoslovakia* in 1968; and the second in the series, which he finished editing during the weeks before his death, appeared in 1971. Dedicated to him, this second volume included his final article, 'Czechoslovak Zionism: Analecta to a History', which surveyed the growth of the movement in which he had taken so prominent a part.

During these last, more relaxed years, Rabinowicz found time not only for his scholarly pursuits but also for painting, which he had long neglected. His lifelong interest was evidenced by his continual collection of works of art, notably by his friend Sir Jacob Epstein. Busts of Churchill and Sholem Asch were his particular pride, and he himself was the subject of one of Epstein's creations. But now he resumed his own work in canvases that recalled his youth, notably scenes from Boskowitz and from the ghetto, portraits of the characters he remembered, and vistas of the surrounding countryside. Soon, however, he turned to a more immediate source of inspiration, the Holy Land. He took his sketchbook wherever he went in Israel, lavishing much care on his special favourite, the Old City of Jerusalem, with its markets and narrow streets. Old Jews, Talmudic scholars, labourers, and members of his own family provided him with subjects until only a few weeks before his death.

Some of the paintings remained unfinished, as did the exhaustive, multi-volume history of Zionism on which he had been engaged for decades. The research had taken him to all the major repositories and archives of modern Jewish history, but he had managed to complete the story only until the year 1920, the eve of his own entry into the movement. His documents and manuscripts for the project have been deposited in the Zionist Archives in Jerusalem, awaiting the future scholars who can complete the task he began.

The vivid memories he left in the minds of those who knew him testify to the boundless energy, the quick and deep concern for others, the kindness and conscientiousness he always displayed. He took his public services seriously, never joining a committee or board without giving of his time unstintingly. And when he was engrossed in a project he gave it a level of concentration that excluded all else. One Hallowe'en evening, when children from the neighbourhood traditionally went round asking for sweets, he was alone in the house working on an article. He became so absorbed in his researches that he never heard the doorbell ring; and the sweets remained in the dish undistributed. Characteristically, he was filled with remorse, and made sure that in subsequent years such responsibility would not fall on him alone.

Always outgoing and humane, he used his creativity to forward momentous causes. His overwhelming commitment to his fellow-Jews took many forms—political activity, propaganda, social work, oratory, and scholarship—but throughout these tireless efforts there was a calm confidence that his Zionist ideals would triumph. Thus, the greatest joy of his life was the establishment of the State of Israel, to whose origins he himself had made so notable a contribution. Yet he was always able to derive satisfaction even from simple occurrences—the discovery of a new document or an interesting Jerusalem street, or just a chance meeting with an old Zionist colleague. It was this richness and variety that gave his life both its momentum and its special quality as a distillation of the very best from that brilliant Central European Jewish culture of which he was one of the last representatives.

INDEX

to Sir Solomon de Medina

Entries printed in italics in this index include the many works which the author consulted and which he quoted from or referred to, thus constituting a **Bibliography**. Most of the titles occur in the footnotes, together with their authors' names, which also appear separately in the index. Among other works the author consulted are *A Dutch Burial-Ground* (Harris), *Encyclopedia Judaica* (Berlin), *Jews in the Canary Islands* (Lucien Wolf), and *Juifs de Bordeaux, leur situation morale et sociale de 1550 à la Révolution* (Georges Cirut, 1920). State Papers, whose identification numbers are regularly given in the footnotes, are not included here because of their great number.

Abbot, Mr., 30
Abrahams, Dudley, 90n, 92n
Acosta—*see* Costa, da
Acts of the Privy Council—Colonial Series III (1720–1745), 88nn
Albemarle, Duke of, 84
Aldag, Peter, 67n
Allen, Tho., 79
Amsterdam, 1, 2, 3, 5, 23, 24, 26, 33, 34, 35, 41, 42, 52, 53, 56, 63, 73, 75, 76, 77, 80, 83nn, 92—*see also* Economic Activities
Anne, Queen, 1n, 20n, 21, 24, 63, 66, 67n—*see also* Queen Anne (Paul)
Annesley, Fra., 65
Antisemitism, 67, 72, 74
Antwerp, 42, 52
Arias, Aron, 57, 97
Army contractors, 14, 15, 16, 26, 27, 28, 31, 32, 33, 35, 37, 38, 43, 54, 67n—*see also* Medina, Sir Solomon de: army contractor
Ascama (pl. Ascamot), 6n, 7, 8, 10, 12, 24, 80—*see also* Libro de los Acuerdos . . .
Athenaeum, The, 71n
Azevedo, Ester d' (Lady [Solomon] de Medina), 3, 5, 58, 86n, 87n, 92—*see also* Medina, Sir Solomon de: wife
Azevedo, Sarah d', 96, 103

Bank of England, 18, 28, 59, 96, 98, 101, 103, 104, 106, 107—*see also* Medina, Sir Solomon de: Bank of England connections

Bank of England, a History (J. H. Clapham, 1944) 28n
Baptismal register entry, 87n, 90
Barnett, Lionel D., 2n, 12n
Barzilay, Ishak, 9
Beaker, Nicho., ship's master, 32
Becher, John, 4
Benson, Robt., 38
Berahel, Jacob—*see* Lix, Francisco de
Bergeyk, Comte de, 44
Berio (Zibezberio), James Zibez, 79
Berlin, 70, 72
Bertie, Hen., 65
Bevis Marks—*see* Spanish and Portuguese Synagogue, London
Bevis Marks Records II (Lionel D. Barnett, ed., 1949), 12n, 83n, 84n, 86n, 89n, 92nn, 93n
Blieck, Jan de, 112
Blockar, Solomon, 79
Bloom, Herbert, 14n, 24n, 34n, 35nn, 36n, 83n
Bols, William de, 110, 112
Booth, Mr., 78
Bordeaux, 1, 2, 77, 82
Bradbury, Geo., 78, 79
Brandon, Ribca, 96, 103
Brandon, Simha (Mrs. Aron Arias), 57, 97
Brenchley, John, 91
Bridegrooms of the Law—*see* Hatanim
Brief Historical Relation of State Affairs from September 1678 to April 1714 (Narcissus Luttrell, 1857), 19n
Brugmans, Hendrik, 15n, 56n
Bryan, John, 79

143

Bryant, Edward Christopher, 91
Brydges, James, 47, 48, 49, 50, 114
Brydges, John, 58
Bueno, Joseph, 79
Bull (?), Joseph, 79
Bull, Peter, 79
Buston, Sir William, 85
Butler, Joseph, 79
Butler, Tho., 76
Byrloam, Simon, 79

Cadogan, General, 45
Calender of the Court Minutes etc. of the East India Company 1674–1676 (Ethel Bruce Sainsbury, 1935), 17n
Campion, H., 65
Capadae (? Capadose) family, 4
Cardonnel, Mr., 63, 114
Cardoso, David Usiel, 57, 97
Cardosso, Phelipi, 44
Carlisle, Lord, 84
Carroen, Moses, 79
Carroon, James, 79
Cartwright, Captain Henry, 50
Carvajal, S^{ra} de, 7
Casquet, William de, 79
Castano, Joshua, 42, 52, 54
Cautivos, 8
Chambers's Encyclopaedia (1950), 27n, 70n
Charles II, King, 2, 10, 20n, 75n
Children—*see* Education
Chrétien contre six juifs, Un (ed. Moland), 72n
Churchill, Sir Winston, 27, 68, 69, 70n
Clapham, J. H., 28n
Clifford, 76
Coal, 23
Cobbett's Parliamentary History of England from the Earliest Period to the year 1803 (1810), 64n, 65nn
Cockrell, Tho., ship's master, 32
Collins, John Churton, 71n
Colonial America and West Indies (S.P.), 79n
Colyer-Fergusson, Sir Thomas, and his *Collection of Jewish Pedigrees*, 90nn
Commodities, trading, 15, 17, 23, 24, 30–34 *passim*, 38–41 *passim*, 43, 44, 45, 47, 48, 53, 64, 65, 67, 69, 70, 78, 79, 81
Conquer, William de, 79

Corea, Antonio, 79
Cory, John, 78
Costa, da (Acosta), banker, 72, 74
Costa, Alvarez (Alvaro) de, 17, 79
Costa, Anthony Mendes da, 73
Costa, Isaac Telles da, 12
Costa, John Mendes da, 56, 73
Costa, Manuel de, 34, 87
Coulrick, Elizabeth (Mrs. Solomon de Medina), 90
Coxe, William, 64n, 65n, 69n
Cundall, Frank, 85n

Daily Journal, 73n
Davila, Isaac Israel, 11
Davis, N. Darnell, 85n
Decisions of the Lords of Council and Session. From June 6, 1678, to July 30, 1712 (Sir John Lander, 1761), 20n
Decree in the Case of Solomon de Medina Mosesson and Company, Merchants in London, and Roderigo Pacheco, Jacob de Lara and Manuel da Costa (New York, 1728), 34n, 87nn
Debts, war—*see* War debts
Denization, 1n, 3, 23, 75, 76, 77, 80, 84
—*see also* Medina, Sir Solomon de: denization
Denizations and Naturalizations (Wm. A. Shaw, 1911), 1n
Depont, John, 79
Dias, Moses b. Isaac, 21n, 116–117
Disraeli, Benjamin, 67
Dodsworth, Christopher, 78–79, 81
Dortha, Ishak Soarez, 9
Dublin, 15n
Dubnow, Simon M., 14n, 71, 73
East India Company, 17, 24n, 37, 51—*see also Calender of the Court Minutes* ...; Medina, Sir Solomon de: East India Company connections
Economic Activities of the Jews in Amsterdam in the Seventeenth and Eighteenth Centuries (Herbert Bloom, 1937), 14n, 83n
Edinburgh Review, 73n
Education, 10–11—*see also* Talmud Torah
Emden, Paul H., 14n
Endenization—*see* Denization
Evans, Stephen, 79
Examiner (newspaper), 66
Exchange rate, 28–29, 31

INDEX TO SIR SOLOMON DE MEDINA 145

Faxaia, Peter de, 79
Fermaco, John, 79
Fernandes Nunes—*see* Nunes
Festskrift i Anledning af Professor David Simonsens (Copenhagen, 1923), 6n
Flanders, 15, 16, 27–31 *passim*, 50, 61
Fletcher, Sir Henry and Lady, 91
Fonseca, Alvaro de, 12
Fonseca, Mrs. Ribka Aboab, 57, 97, 101
France, 61, 63, 66, 76, 77—*see also* Bordeaux
Francia, Abraham Rodriz de, 8, 9
Francia, Isaac Israel de, 7
Francia, Simon, 79
Francisco, Jasper, 79
Frank, Abraham, 15n, 56n
Friedenberg, Albert, 85n
Friedman, Lee M., 14n

Gabay (Treasurer), 8, 9, 85—*see also* Medina, Sir Solomon de: Gabay
Gambling, 88, 106, 107
Garo, Simon, 79
Gaster, Moses (Haham), 9n, 12nn, 84n
Gentleman's Magazine, 87n, 89n, 90nn, 91n
George, M. Dorothy, 10n
Geschiedenis des Joden in Nederland (Brugmans and Frank, Amsterdam, 1940), 15n, 56n
Gideon, Rowland, 13
Giuseppi, J. A., 18nn
Godolphin, Earl of, 46, 63, 70
Gomez, Isaac, 79
Gomez Sera—*see* Serra
Gonzalez, Jacques, 79
Goodman, Alexander, 79
Goods—*see* Commodities
Graetz, Heinrich 71, 72, 73
Guildhall Records, 3
Guydett, Mr., 16

Hague, The—*see* The Hague
Haham, 11
Hampton Court, ix, 19, 20
Harley, E., 40, 55, 69
Harley, Robert—*see* Oxford, Earl of
Harris, Peter, 79
Harris, William, 79
Harwood, John, ship's master, 32
Hatanim (Hatan Bereshith; Hatan Torah), 6, 7, 8, 77—*see also* Medina, Sir Solomon de: Hatan Bereshith

Haverfield, F., 91
Haverfield, Rev. Thomas Tunstall, 91
Headland, Dr. Edward, 90–91
Hennquir, Peeter, jun., 115
Henriques, Jacob Lopez, 86n
Henriques, Jacques, 35n
Henriquez, Peir, 79
Henriquez, Peter, 79
Hettier, M., 72
Hill, Mr., 29
Hirschl, Abraham, 70, 72
History of the Ancient Synagogue of the Spanish and Portuguese Jews (Moses Gaster, 1901), 9n
History of the Jews (Simon Dubnow), 73n
History of the Jews (Heinrich Graetz), 71n, 72
History of the Jews in England (Cecil Roth, 1941), 36n, 75n
History of the Marranos (Cecil Roth, 1941), 86
Honen Dal (The Hague), 56, 95
Holland, 2, 4, 13, 14, 15, 80, 82, 84, 86
status of Jews, 14
—*see also Hoofdstukken uit de Geschiedenis...; Geschiedenis des Joden...* ; Low Countries
Hoofdstukken uit de Geschiedenis der Joden in Nederland (Jacob Zwarts, 1929), 83n
Hutten (Hutton), Gerrit, 103, 110

Imposta, 7, 8, 10, 24–25, 71—*see also* Medina, Sir Solomon de: imposta
Ince, Brigadier Cecil Edward Ronald, 70n
Inchiquin, Lord, 84
India Office, 17n, 18nn, 42n
Intelligence, military—*see* Medina, Sir Solomon de: express message system
Interest rate, 28
Ireland, 62
Isaac, Mordecai, 79

Jackson, Peter, 79
Jackson, Philip, 4
Jamaica, naturalisation, 84–85
James II, King, 10, 20n
Jennings, Sarah, Duchess of Marlborough, 73
Jephson, William, 15, 28

Jepson, Mr., 79
Jewish Chronicle, 1n, 67n
Jewish Encyclopaedia (USA), 14n
Jewish Historical Society of England, vii—*see also* Miscellanies (*JHSE*); Transactions (*JHSE*)
Jewish World, 1n
Jews of Britain (Paul H. Emden, 1944), 14n
Jews in the Literature of England (Montague F. Modder, 1944), 14n
Johnson, John, 79
Jong, Joh. de, 15n
Joseph de Medina & Sons (Amsterdam), 24, 26, 33, 34, 35, 42, 51, 52, 57, 58, 59, 73, 80, 83, 92, 96
Journal to Stella (Jonathan Swift, 1712), 67n
Juden und Das Wirtschaftsleben (Werner Sombart, 1928, c. 1911, 1918), 36n
Juden Erobern England (Peter Aldag, 1940), 67n

Kingdon, Lemuell, 15
Knighthood, ix, 22—*see also* Medina, Sir Solomon de: knighthood
Knights of England (W. A. Shaw, 1906), 20n

Lamago (Lamego), Jacob de Aharon, 92
Lamago, Rachel—*see* Medina, Rachel de (daughter of Moses)
Lander, Sir John, 20n
Lansdowne, 34n
Lanson, Gustave, 72, 73
Lapo, George de, 79
Lara, Jacob de, 34n, 87n
Laski, Neville, 9n
Laws and Charities of the Spanish and Portuguese Congregation of London (Neville Laski, 1952), 9n
Lean, John de, 79
Lechmer, N., 78
Leghorn, 75
Leneve, Captain (Customs officer), 78, 79
Le Neve, Peter, 20
Levi, Joseph, 79
Levi, Pas., 79
Levi, Solomon, 79
Levy, Michal, 78
Libro de los Acuerdos: being the record and accompts of the Spanish and Portuguese Synagogue from 1663–1681, El (trans. L. D. Barnett, 1931), 2n, 6nn, 8–12, 24n, 77—*see also* Ascama; Laws and Charities . . .
Lisbon, 34
Lister, Theo., 65
Litigation, 86n, 87, 88n—*see also* Medina, Sir Solomon de: will, litigation
Little London Directory of 1677, 3, 23
Lix, Francis de (Jacob Berahel), 76, 77
Lockhart, George, 64, 65
London Gazette, 19, 73
London Life in the XVIIIth Century (M. Dorothy George, 1925), 10n
Lopez, Andrew (Abraham), 84
Lopez family, 1n
Lords Commissioners of the Treasury, 17, 23, 30, 31, 48, 49, 56, 69, 78, 79, 81
Lords in Council and Session (Scotland), 20, 21n
Lottery, Million Act, 17—*see also* Medina, Sir Solomon de: Commissioner for the Million Act Lottery
Louis XIV (France), 15, 61
Low Countries, ix, 64, 65, 66—*see also* Flanders; Holland
Lowndes, William, 79
Luttrell, Narcissus, 19n
Lynch, Sir Thomas, 84

Machado, Jacob Hiskia, 41, 53, 58
Machado, (Moses) Anthony Alvarez, 15, 33, 37, 38, 64, 65, 68
Machado and Pereira (Holland), 14–15, 16, 26, 27, 31, 38, 40, 42, 61, 62—*see also* Medina, Sir Solomon de: Machado and Pereira
Mahamad, 7–10 *passim*, 85
Malez, Josias, 79
Mangold, Wilhelm, 70n
Mansell, Sir Thos., 38
Marlborough, Duke of, ix, 3, 16n, 27, 38, 43, 44, 49, 53, 54, 61–70, 73—*see also* Jennings, Sarah, Duchess of Marlborough; Medina, Sir Solomon de: Marlborough
Marlborough Dispatches (ed. Sir George Murray, 1845), 3n, 53n, 62n, 68n, 69n

Marlborough, His Life and Times (Sir Winston Churchill, 1933-1938), 27n
Marques, Antonio Rodriguez, 79
Marques, Diego Rodriguez, 82
Marques, Joseph, 79
Marranos 1, 86—*see also History of the Marranos*
Marshall, George W., 20n
Mary, Queen (wife of William III), 20n
Mawton, Mary (Mrs. Ferdinand de Medina), 90
Medina (Holland), 73
Medina, Aaron de (son of Isaac, sen.), 83, 96, 103
Medina, Aaron de (son of Moses), 87, 89
Medina, Abigail de (daughter of Moses; Mrs. Isaac de Abraham Haim Mendes), 87, 89
Medina, Abraham de (son of Moses), 58, 59, 87-90 *passim*, 96, 100, 102, 104, 107, 108
Medina, Amelia de (Mrs. Henry Walker), 91
Medina, Caroline Sophie (1st, Mrs. Edward Christopher Bryant; 2nd, Mrs. John Thomas Tunstall Haverfield), 91
Medina, Charles Smith de, 91
Medina, Deborah de (daughter of Joseph Aaron de Medina; Mrs. Abraham de Solomon de Medina), 84
Medina, Deborah de (daughter of Sir Solomon; Mrs. Moses de Medina)—*see* Medina, Sir Solomon de: daughter
Medina, Diego de, 2, 3, 23, 75, 76, 79—*see also* Medina, Sir Solomon de: Diego, identity with
Medina, Elias de, 79
Medina, Eliza de (Mrs. Edward Headland), 90
Medina, Elizabeth de—*see* Coulrick, Elizabeth
Medina, Elizabeth de (daughter of Solomon and Elizabeth), 91
Medina, Ester de (daughter of Moses; Mrs. Solomon alias Francisco de Medina), 34, 57, 87, 92, 99, 101
Medina, Ester de (wife of Isaac, sen.) —*see* Miranda, Ester Nunes
Medina, Ester de (wife of Sir Solomon) —*see* Azevedo, Ester d'; Medina, Sir Solomon de: wife
Medina, Ferdinand de (son of Ferdinand), 91
Medina, Ferdinand de (son of Solomon and Elizabeth), 90, 91
Medina, Francesco de (father of Sir Solomon), 1, 2, 92
Medina, Francesco alias Solomon Hiskia de—*see* Medina, Solomon Hiskia alias Francesco de
Medina, Gracia de (Gracia Pereira)—*see* Medina, Sir Solomon de: mother
Medina, Isaac de, junior (son of Moses; nephew of Isaac, senior), 83, 87, 89, 94, 96, 100, 102, 104, 107, 108
Medina, Isaac de, senior, 33, 34, 59, 83, 84, 85, 95, 96
Medina, Jochebeth de (daughter of Moses), 57, 59, 87, 93, 99, 101, 109
Medina (Joseph) & Sons—*see* Joseph de Medina & Sons
Medina, Joseph de (son of Abraham), 58, 89
Medina, Joseph Aaron de (brother of Sir Solomon), 1, 24n, 33, 34, 82-93, 114—*see also* Joseph de Medina & Sons
Medina, Louisa Charlotte de (Mrs. John Brenchley), 91
Medina, Marianne de (Mrs. John Swete), 91
Medina, Mary de—*see* Mawton, Mary
Medina, Moses de (son of Abraham), 89
Medina, Moses de (son of Joseph), 3, 5, 24, 33, 34, 35, 58, 59, 82, 84-87 *passim*, 89, 92, 93n, 99, 101, 102, 108, 109—*see also* Medina, Sir Solomon de: son-in-law
Medina, Rachel de (daughter of Isaac, sen.; wife of Solomon alias Diego), 34, 59, 83, 88, 89, 100, 104, 105, 106
Medina, Rachel de (daughter of Moses; Mrs. Jacob de Aharon Lamago), 57, 83, 87, 92, 99, 101, 103
Medina, Rachel de (wife of Joseph) 34n, 82, 83, 84
Medina, Ribca de (daughter of Moses), 87, 92-93

INDEX TO SIR SOLOMON DE MEDINA

Medina, Sarah de (daughter of Abraham), 58, 89
Medina, Sarah de (daughter of Solomon and Elizabeth), 91
Medina, Sarah de (wife of Abraham)— *see* Sequeira, Sarah de Joseph Israel
Medina, Sir Solomon (Selomoh) de:
  army contractor, ix, 14, 16, 26, 30–33 *passim*, 37–42, 53–56, 113–115— *see also* financier; Marlborough, Duke of, connections with
  Bank of England connections, 18
  bankruptcy—*see* Voltaire
  birth date and place, 1, 2
  brother—*see* Medina, Joseph Aaron de (main entry)
  coat of arms, 20
  Commissioner for the Million Act Lottery, 17
  daughter (Deborah, Mrs. Moses de Medina), 3, 32, 34, 50, 59, 82, 83, 84, 87, 89, 92, 98, 99, 101, 104, 106, 113
  death, 4
  denization, 2, 3, 75, 77, 80
  Diego, identity with, 1, 75–81, 87
  domicile, 1–4 *passim*, 13, 14, 25, 26, 33, 36, 55, 63, 73, 80, 84, 94
  East India Company connections, 17–18, 23, 42
  Elder of Congregation, 12
  express messages system, 35, 36, 38, 39, 40—*see also* Secret Service
  family, 1–5
  father (Francesco), 1, 92
  financier, ix, 16, 17, 25, 26–36—*see also* army contractor; loans to Government
  Gabay of congregation, 8–11 *passim*
  gifts to synagogue, 7, 8
  Hatan Bereshith, 6, 7, 8, 77, 80
  history, early, 1–5, 20, 75, 80
  home, Richmond Hill, 3, 4, 19
  Imposta, 1, 6, 12, 24, 25, 77, 80
  Joseph de Medina & Sons—*see* main entry
  knighthood, ix, 20, 21–22, 116
  loans to Government, 16, 17, 26–33, 40, 43—*see also* financier
  losses—*see* Marlborough, connections with: wealth and decline
  Machado and Pereira, connections with, 14, 15, 16, 19, 26–30 *passim*, 33, 61
  Marlborough, Duke of, connections with, 61–70, 73
  marriage—*see* wife
  merchant, 18–19, 23–25, 44—*see also* financier
  mother (Gracia Pereira), 1, 16
  parents—*see* father; mother
  Parnas, 9, 86
  position in Jewish community, 6–13
  prominence in public life, ix, 14–22
  Secret Service connection, 16—*see also* express messages system
  ships—*see* Ships, main entry
  signature, facsimile, 11, 12n
  son-in-law and nephew (Moses), 3, 4, 18, 23, 24, 26, 32–35 *passim*, 37, 38, 42, 46, 48–52 *passim*, 55, 62, 67, 68, 69, 83, 84, 94, 95, 97–100 *passim*, 104, 106, 107, 112, 113–115
  Stock Exchange speculation, 35, 36, 38, 40
  synagogue connections, 1, 2, 6–13, 21, 24—*see also* Elder; Gabay; Hatan Bereshith; Imposta; Parnas
  tombstone, 5
  turnover, 25, 44
  and Voltaire—*see* Voltaire, main entry
  wealth and decline, 43–60
  wife (Ester d'Azevedo), 3, 5, 57, 59, 94–112—*see also* Azevedo, Ester d', main entry
  will, 3n, 4, 18, 34n, 51n, 52, 56–60, 73n, 82, 83n, 87n, 88, 92, 94–107, 108, 110, 111
  will litigation, 58, 59, 60
  William III, relations with, 4, 14, 22
Medina, Solomon de (son of Abraham), 90
Medina, Solomon de (son of Solomon and Elizabeth), 91
Medina, Solomon alias Diego de (otherwise Solomon de Medina Mosesson = de Moses), 8, 34, 59, 80, 83, 87, 88, 89, 96, 99–102 *passim*, 104, 105, 106, 108
Medina, Solomon Hiskia alias Francesco de, 5, 33–34, 53, 58, 59, 92, 94, 95–96, 99, 109, 110, 111
Medina, Solomon de Moses de—*see* Medina, Solomon alias Diego de; Mosesson

INDEX TO SIR SOLOMON DE MEDINA 149

Medina, Sophie de—*see* Normansell, Sophie
Meditaciones Sobre la Historia Sagrada de Genesis (Moses Dias, 1697; 1705), 117
Mementon, Abraham, 57, 97
Mementon, Isaac, 57, 97
Mementon, Sarah (Mrs. David Usiel Cardoso), 57, 97
Mémoirs de l'Académie des Sciences, Arts et Belles-Lettres de Caen (M. Hettier, Caen, 1905), 72n
Memoirs of John, Duke of Marlborough, with his Original Correspondence (William Coxe, 1820), 64n
Mendes, Abigail, 57, 99, 101
Mendes, Abraham, 93, 99
Mendes, Abraham de Solomon, 84
Mendes, Isaac de Abraham Haim, 93
Mendes, Isaac de Joseph Yessurun, 12
Mendes, Moses, 93
Mendes da Costa—*see* Costa
Miez, Joseph, 79
Million Act Lottery—*see* Lottery, Million Act
Miranda, Ester Nunes (Mrs. Isaac de Medina, sen.), 83
Miranda, Jacob (Jeronimo Fernandez) de, 6, 7, 8, 76, 77
Miscellanies (JHSE), 84n, 90n
Modder, Montague F., 14n, 66n
Moland, L., 73n
Monatsschrift fuer die Geschichte und Wissenschaft des Judentums, 72n
Monmouth, Duke of, 15, 78
Monsanto, David Roiz, 57, 97, 101, 111
Montefiore, Sir Moses, ix, 22
Morayquyn, Thomas, 107
More, Thomas (Deputy Paymaster), 56
Morris, Samuel, ship's master, 31
Morris, Thomas, ship's master, 32
Mosesson, Solomon de—*see* Medina, Solomon alias Diego de
Mosesson and Company, 34n
Murray, Sir George, 3n
Muyssart, Abraham, 52

Nash, Walter, 79
Navy Commissioners, 15, 28
Nederlandsche Israelitische Weekblad, 15n
Netherlands—*see* Holland
Neve—*see* Le Neve
'New Christians'—*see* Marranos

Newman, Peter, 79
Nieto, Haham David, 12, 85
Nieuw Israelietisch Weekblad, 4n, 92n
Normansell, Sophie (Mrs. Solomon de Medina), 91
Nunes, Ester de Abram Fernandes, 12
Nunes, Isaac Fernandez, sen., 4
Nunes Torres, Rabbi David, 57, 97, 98, 100, 103, 107

Œuvres complètes (Voltaire; ed. L. Moland), 73n
Oldest Printed List of Merchants and Bankers of London (reprint 1863), 3n
Oliveira, Abraham de, 7, 9
Orange, House of, 14—*see also* William III, King
Oxford and Mortimer, Earl of (Robert Harley), 38, 51, 69, 113

Pacheco, Roderigo, 34n, 87n
Paget, Henry, 38
Paiva, Isaac de, 8, 9
Palmer, John, 79
Papez, Joseph, 79
Parnas (Warden), 8, 9, 86
refusal to accept office, 9 *passim*, 86
—*see also* Medina, Sir Solomon de: Parnas
Partition Book, 20, 21
Paul, Herbert W., 70n
Pedigrees, Collection of Jewish (Sir Thomas Colyer-Fergusson), 90nn
Pedigrees of the Knights (made by King Charles II, King James II, King William III and Queen Mary, King William alone, and Queen Anne) (Peter Le Neve, ed. George W. Marshall, 1873), 20
Pendegrass, Monsieur, 94
Penso, David, 12
Pepall, Caleb, 79
Pereira—*see* Machado and Pereira
Pereira, Gracia—*see* Medina, Sir Solomon de: mother
Pereira, Isaac de, 62
Pereira, Juda, 53, 54
Pereira, Manuell Lopez, 1
Peres, Daniel, 12
Perrera, Peter, 79
Perrero, Jasper, 79
Phillips, John, 79
Picciotto, James, 1

Pomea, Oder, 79
Population, Jewish, 78
Portugal, 86
Portugees-Israelitische Gemeente, Amsterdam, 5
Powlet, John Earle, 38
Proceedings of the American Jewish Historical Society, 14n, 85n
Promesa, 8
Providiteurs, 15, 61, 62—*see also* Medina, Sir Solomon de: Army contractor
Pryer, Alexander, 79

Queen Anne (Herbert W. Paul, 1906), 70n
Questions sur l'Encyclopédie (Voltaire), 71n

Ramsay, Peter, 79
Ranelagh, Richard, Earl of, 15, 28
Revue Latine, La, 72n
Richardson, Nicholas, ship's master, 32
Richmond, 3, 4, 19, 86
Robertson, William, 79
Rodriguez, Alphonso, 79
Rodriguez, Antonio, 79
Roth, Cecil, 36, 75, 80, 86n
Rothschilds, 36
Royal Exchange, 86
Ryswick, peace of, 19

Sainsbury, Ethel Bruce, 17
Samuel, Wilfred S., 13n, 75, 87n
Sarmento, Jacob de Castro, 86
Sasportas, Haham Jacob, 11
Sasportas, Samuel, 11
Savoy, Duke of, 30
Schools—*see* Education
Scopens, John, 79
Seeligmann, Sigmund, 7n
Sellion, 79
Sequeira, Sarah de Joseph Israel (Mrs. Abraham de Medina), 89, 90
Serra, Antonio (Jacob) Gomez, 76, 79, 84
Seymour, James, sen. (?), 79
Shaw, Wm. A., 1n, 20n
Shephard, Nathanel, 4
Shephard, William, 4
Sherrin, John, 79
Shillman, Bernard, 15n
Shippen, Will., 65

Ships: *Brother's Desire*, 32; *Constant*, 31; *Elizabeth*, 32; *Industry*, 32; *John*, 32; *Soes-Dyke*, 78; *Success*, 32
Sickerstaffe, C., 76
Sieveking, A. Forbes, 71, 72
Silva, Haham Joshua de, 11
Silver trading, 78–79, 81
Simonsen, David, 6n
Sketches of Anglo-Jewish History (James Picciotto, 1875), 1n
Sloper, Mr., 48, 49
Sluys, D. M., 56
Smith, George, 79
Smith, M. F. J., 35n
Snelling, Will., 79
Solomons, Israel, 85nn
Sombart, Werner, 36
South Sea Company, 52, 53, 57, 59, 88, 97, 105, 106, 108, 109
Spanish and Portuguese Synagogue, Amsterdam, 56, 95, 96, 104
Spanish and Portuguese Synagogue, London, ix, 1, 6–13, 57, 73, 86, 96; *Minute Book* 9n, 12nn; office refused 9, 86; reforms 9–13; wedding, first, 12—*see also* Ascama; *History of the Ancient Synagogue* . . .; *Libro de los Acuerdos* . . .; Medina, Sir Solomon de: synagogue connections
Spanish and Portuguese synagogue, The Hague—*see* Honen Dal
Spanish Succession, War of—*see* War of the Spanish Succession
'Species hollandia judaica', 6
Speculation, 35, 36—*see also* Medina, Sir Solomon de: stock exchange, speculation
Spender, Harold, 91
Spender, J. A., 91
Spender, Stephen, 91
Spinoza, 85
Stevens, Captain Francis, 55, 56
Stock Exchange—*see* Medina, Sir Solomon de: stock exchange; Royal Exchange
Stone, Antho., 79
Strachey, Lytton, 73n
Swann, Lawrence (artist), 78, 79
Sweet, Mr., 48, 55, 63, 64
Sweetaple, John, 79
Swete, Professor Henry Barclay, 91
Swete, Rev. John, 91

INDEX TO SIR SOLOMON DE MEDINA 151

Swift, Jonathan, 66n, 67
Sybil (Benjamin Disraeli), 67n
Symons, Derrick, 79
Sythoff, John, 94, 103, 107, 110, 111, 112

Talmud Torah, 11, 56, 95, 96
Telles da Costa—*see* Costa
Terra Santa, 8
The Hague, 56, 63, 65, 69, 73, 94
Thiriot, 71, 72
Thompson, John, 79
Tijd-Affaires in Effekten aan de Amsterdamsche Beurs (M. F. J. Smith, 1919), 35n
Tombstones (Solomon and Ester de Medina)—*see* Medina, Sir Solomon de: tombstone
Torres, Nunes—*see* Nunes Torres
Transactions of the Jewish Historical Society of England, 13n, 15n, 20n, 85n
Treasurer of the Spanish and Portuguese Congregation—*see* Gabay
Treby, George, 78

Universal Jewish Encyclopaedia (USA, 1942), 14n
Utrechtsche Compagnie, 83, 88, 92

Vallentine, John, 79
Van Alphen, Lambert, 103, 107, 110, 112
Vanderboon, Geo., 79
Vanderhorne, John, 79
Vanderkaa, Adrian, 42, 52, 53, 54
Vanderkaa, Leonard, 41
Vanderpost(?), John, 79
Van Dillen, J. G., 15nn, 34n, 52n
Van Hatteren, Hendrick, 53
Vanhine, John, 79
Van Limburch, Frans, 57, 98

Van Neck, Joshua, 53
Van Tangeren, Bernard, 52, 53
Vanvolgli, G., 79
Vaughan, Lord, 84
Veiga, Jeromino Soarez de, 111
Venneck, Gerard, 53
Vernon, Thomas, 54
Voltaire, 61, 70–74
Voltaire in England (John Churton Collins, New York, 1886), 71n
Voltaire's Rechtsstreit mit dem koeniglichen Schutzjuden Hirschel, 1751 (Wilhelm Mangold, Berlin, 1905), 70n

Walker, Henry, 91
War debts, 19—*see also* Medina, Sir Solomon de: financier
War of the Spanish Succession, ix, 33, 62, 68
Warden of the Spanish and Portuguese Synagogue—*see* Parnas
Weltgeschichte des judischen Volkes (Simon M. Dubnow, 1925–1929), 14n
White, Th., 79
William III, King, ix, 4, 10, 14–22, 26–31 *passim*, 33, 61, 71, 80, 85— *see also* Medina, Sir Solomon de: William III, relations with
William of Orange—*see* William III, King
Winnington, S., 65
Witness, Jew as, 21
Wolf, Lucien, 1, 15n, 20n, 26, 62n, 67n, 70, 71, 75, 76, 77
Worden, Sir John, 78, 79
Wright, Mr. (Customs officer), 78, 79

Zarfatti, Joshua, 85
Zevi of Altona, Haham, 85
Zibezberio, James, 79
Zwarts, Jacob, 4, 15nn, 38n, 83n, 88n, 92nn

J.M.S.

INDEX

to Oskar K. Rabinowicz

American Friends of the Hebrew University, 135
American Israel Cultural Foundation, 135
American Zionist, 135
Anglo-Federal Banking Corporation, 128, 130
Arabs—*see* Rabinowicz, Oskar: Arabs, displaced
Asch, Sholem, 141
Aspern, 121
Association of Jewish Journalists and Authors, 131
Austria, 127—*see also* Vienna

Balfour Declaration, 122, 133–136 *passim*
Bar Ilan University, 139
Behr, Jan, 140
Beloff, Professor Max, 131
Ben Gurion, David, 132, 138
Berlin, Sir Isaiah, 131
Bialik, Nachman, 131
Blau-Weiss (Blue-White) organisation, 126
B'nai B'rith, 134
Boskowitz, 121, 141
Brandeis, Justice, 133
Brandeis University, 135
British Friends of the Midrashia, 131
British Museum, 138
Brno, 121, 122, 123, 139
Brodie, Rabbi Sir Israel, 131

Carlsbad, 122
Carmel College, 131
Centre of Jewish Documentation, 131
Chamberlain, Neville, 132
Churchill, Sir Winston—*see* Rabinowicz, Oskar: Churchill, Sir Winston
Chvalkovsky, Frantisek, 127
Clements Trading Company, 128, 131

Committee to Boycott Nazi Germany, 125, 127
Committee on the Restoration of Continental Jewish Museums, Libraries, and Archives, 129
Committee for Zionist Research, 129
Conference on Jewish Social Studies (USA), 135
Congress Journal, 122
Conversion, 123, 130, 139
Council of Christians and Jews, 129
Cromwell, Oliver, 132
Cyprus, 136–137
Czechoslovakia, 123–128 *passim*, 139, 140—*see also* Jews of Czechoslovakia . . .; Relief Committee for Jews from Czechoslovakia; Society for the History of Czechoslovak Jews; *Zeitschrift* . . .

Dead Sea Scrolls, 138

East Africa Scheme, 133, 134, 136
Emigration, Jewish, 127
Encyclopedia Judaica (Jerusalem), 135
Epstein, Sir Jacob, 141
Essays presented to Chief Rabbi Israel Brodie on the Occasion of his 70th Birthday (1967), 139

Fifty Years of Zionism: A Historical Analysis of Dr. Weizmann's 'Trial and Error' (Oskar Rabinowicz, 1950), 132, 133, 136
Fraenkel, Josef, 129
Frank, Jacob, 139
Friends of the Haifa Technion, 138
Friends of the Hebrew University, 131

Gaster, Haham Moses, 133
Germany, 129—*see also* Berlin; Nazism
Gestapo, 127, 128
Greenberg, Leopold, 133

Haifa—*see* Friends of the Haifa Technion
Hampstead Literary and Debating Society, 134
Hebrew University, 131, 137, 138—*see also* American Friends of the Hebrew University; Friends of the Hebrew University
Herzl, Theodor, 122, 126, 129, 133, 135, 136, 137, 139
Herzl, Architect of the Balfour Declaration (Oskar Rabinowicz, 1958), 135
Hickl, Max, 122
Hillel House, London, 131
History of the Jews in Aussee (Moravia) (Oskar Rabinowicz, 1927), 123
Hitler, Adolf, 127
Horovitz, Bela, 128

Illegal immigration—*see* Immigration, illegal
Immigration, illegal, 125
Introduction to the Problems of Ritual Slaughter (Oskar Rabinowicz, 1937), 126
Ish Shalom, Meir, 131
Israel, State of, 130, 133, 137, 138, 139, 141, 142—*see also* Rabinowicz, Oskar: Jerusalem

Jabotinsky, Vladimir, 123, 124, 125, 129, 130, 133
Jerusalem—*see* Rabinowicz, Oskar: Jerusalem
Jewish Agency, 125
Jewish Chronicle, 132, 134n
Jewish Cultural Reconstruction (New York), 129
Jewish Cyprus Project: Davis Trietsch's Colonization Scheme (Oskar Rabinowicz, 1962), 136
Jewish Historical Society of England, 129, 132, 139—*see also Transactions* (*JHSE*)
Jewish Historical Society of Israel, 138
Jewish Legion (World War I), 125
Jewish Memorial Council, 131
Jewish National Home, 134
Jewish Publication Society (USA), 135
Jewish Quarterly Review, 139
Jewish Record Office, 131
Jewish State, 124, 126, 134, 136, 137
Jewish State Party, 125, 130—*see also*

New Zionist Organisation
Jewish Theological Seminary (New York), 135
Jews' College, 131
Jews of Czechoslovakia: Historical Studies and Surveys, 136n, 140
Judenstaat—Medina Ivrit, Der—*see Medina Ivrit*
Judaism, 126, 139, 140—*see also* Rabinowicz, Oskar: Orthodoxy
Jüdische Volksstimme (Brno), 122

Landau Prize, 138
Lazarus, Dayan Harris, 131
Leeds Home for Aged Jews and Home of Rest, 129n
Leftwich, Joseph, 129, 131
London, 128—*see also* Rabinowicz, Oskar: London
Louis XIV, 132

Machover, Dr. J. M., 131
Maimonides, 126
Mansoor, Professor M. (Wisconsin University), 138, 139
Marienbad, 121
Marlborough, Duke of, 132
Maspetiol, Vivian, 133
Medina, Sir Solomon de, 132
Medina Ivrit, 126
Midrashia (Israel), 131—*see also* British Friends of the Midrashia

Nazism, 125, 126, 127, 129—*see also* Gestapo; Hitler, Adolf
New Zionist Organisation, 125, 130—*see also* Jewish State Party; Union of Zionist Revisionists
Nordau, Max, 133

Oliner, Rosa (Mrs. Oskar Rabinowicz) —*see* Rabinowicz, Oskar: wife
Oxford University Jewish Society, 125n

Palestine, 125, 128, 134, 136, 137
Pariser Tagesblatt, 127
Phaidon Press, 128
Prague, 122, 124, 127—*see also* Rabinowicz, Oskar: Prague

Rabb, Professor Theodore K., vii, 121, 136n—*see also* Rabinowicz, Oskar: son

Rabinowicz, Judith K. (Mrs. J. Tapiero, daughter of Oskar Rabinowicz)—see Tapiero, Judith K.
Rabinowicz, Kurt (brother of Oskar), 128
Rabinowicz, Martha (sister of Oskar), 121
Rabinowicz, Oskar K:
  Arabs, displaced, 137
  artist, 135, 140–141
  Berlin, 124
  business, 124, 128, 130, 135
  Churchill, Sir Winston, 128, 129, 132, 134, 139, 141
  citizenship, 128
  collector, 141
  cultural work, 129–132 passim, 135, 137, 138, 139
  Czechoslovakia—see Prague, see also Society for the History of Czechoslovak Jews, main entry
  daughter—see Tapiero, Judith K., main entry
  doctorate, 122, 123
  family, 121, 128
  father (Yehuda), 121, 131—see also parents
  Harrogate, 130
  Jerusalem, 137, 138, 141, 142
  journalist and publicist, 122–127 passim, 132, 135, 136, 138
  library, 131, 132
  London, 123, 124, 127–130 passim
  musician, 121
  Nazism (escape from), 127; (fight against), 125–128 passim
  New York, 135
  Orthodoxy, 124, 126
  Oxford, 128
  parents, 128—see also father
  Prague, 127, 128
  publications, 123, 125, 130, 132–136 passim, 138 (see various titles in index)
  rabbinate, 124
  refugees, organisation of, 125, 127, 137
  Revisionism, 123–127 passim, 130
  soldier and army chaplain, 124, 127
  son (Theodore), 127—see also Rabb, Theodore, main entry
  teaching, 122, 123
  university, 122, 123
  wife (Rosa Oliner), vii, 127
  Zionism, 122–126 passim, 130, 132, 133, 139–142 passim—see also Revisionism
Rabinowicz, Rosa (sister of Oskar), 121, 128
Rabinowicz, Rosa (née Oliner, wife of Oskar)—see Rabinowicz, Oskar: wife
Rabinowicz, Yehuda (father of Oskar) —see Rabinowicz, Oskar: father
Rassco (Israel), 138
Refugees, 125, 128, 129, 130—see also Rabinowicz, Oskar: refugees, organisation of
Relief Committee for Jews from Czechoslovakia, 129
Remember the Days (ed. J. M. Shaftesley: JHSE, 1966), 139
Resettlement tercentenary, 132
Revisionism, 122, 124, 125—see also Jewish State Party; New Zionist Organisation; Union of Zionist Revisionists
Roosevelt, Eleanor, 134
Roosevelt Archives, 134–135
Roth, Cecil, 128, 129, 131, 139
Rothschild, Lord, 132

Sadegorah Rebbe, 121
Sapir, Pinchas, 130
Selbstwehr, 123
Semple, Lord, 130
Shabbatai Zvi, 139
Shapira, M. W., 138
  scroll forgery, 138, 139
Shechita, 126
Sinai Peninsula, 139
Society for the History of Czechoslovak Jews (USA), 135, 140
Sokolow, Nahum, 122, 133, 136
Spinoza's 'God' in the Light of Jewish Religio-Philosophical Sources (Oskar Rabinowicz, 1925), 123

Tagesblatt, 123
Tapiero, Judith K. (née Rabinowicz), vii, 121, 130
Temple Israel Synagogue, White Plains (USA), 137
Toynbee, Arnold, 140
Transactions of the Jewish Historical Society of England, 139

Trial and Error (Chaim Weizmann, 1949), 132, 133
Trietsch, Davis, 136–137

Union of Zionist Revisionists, 123—*see also* Jewish State Party; New Zionist Organisation
United States—*see* Rabinowicz, Oskar: New York
Ussishkin, Menahem, 133

Vienna, 121
Vladimir Jabotinsky's Conception of a Nation (Oskar Rabinowicz, 1946), 130

Wanstead and Woodford Synagogue, 132
Weizmann, Chaim, 122, 124, 125, 132, 133, 136
Weizmann Institute (Israel), 138

When Nations Awake (Oskar Rabinowicz, 1934), 125
Winston Churchill on Jewish Problems: A Half-Century Survey (Oskar Rabinowicz, 1956), 134
Wolffsohn, David, 133
Wolfson, Sir Isaac, 130, 131
World Jewish Congress, 135
World War I, 125, 140
World War II, 127, 128

Zeitschrift für die Geschichte der Juden in der Tschechoslowakei, 123
Zidovske Zpravy, 123
Zionism, 129, 132, 133, 134, 136, 137, 139—*see also* Rabinowicz, Oskar: Zionism
Zionist Actions Committee, 125
Zionist Archives, Jerusalem, 141
Zionist Congress, 122, 123
Zionist Organisation, 125, 133

J.M.S.